# SOLAR WATER PUMPING
## A Handbook

Jeff Kenna and Bill Gillett

**PRACTICAL ACTION**
Publishing

**Intermediate Technology Publications**

Practical Action Publishing Ltd
The Schumacher Centre
Bourton on Dunsmore, Rugby,
Warwickshire CV23 9QZ, UK
www.practicalactionpublishing.org

© Intermediate Technology Publications 1984 and 1985


First published 1984 and 1985\Digitised 2013


ISBN 10: 0 94668 890 7

ISBN 13: 9780946688906

ISBN Library Ebook: 9781780443430

Book DOI: http://dx.doi.org/10.3362/9781780443430




Since 1974, Practical Action Publishing (formerly Intermediate Technology Publications and
ITDG Publishing) has published and disseminated books and information in support of
international development work throughout the world. Practical Action Publishing is a trading
name of Practical Action Publishing Ltd (Company Reg. No. 1159018), the wholly owned
publishing company of Practical Action. Practical Action Publishing trades only in support of
its parent charity objectives and any profits are covenanted back to Practical Action (Charity
Reg. No. 247257, Group VAT Registration No. 880 9924 76).

A great deal of research has been attracted to the possibility that solar powered pumps can help villagers and farmers to tap adequate water supplies. Indeed, the need is acute: two billion people do not have adequate domestic water, even though it is a primary requirement for health, diet, and agriculture.

So far, the conceptual recognition that solar water pumps might fill the void has not led to concrete achievements on a scale commensurate with the challenge at hand. However, as a result of intensive work at an international level, solar pumping technology is now viable. This book describes the technology, and most importantly it shows that there are some conditions under which solar pumps already can provide the best solution to local water needs. Furthermore, it quantifies these conditions and it offers a methodology which water supply specialists can use to compare and evaluate available pumping options.

The reader is led step by step through the necessary analyses, including determination of pump requirements, specification of solar pump performance, and comparison of economic data. As a result he or she can obtain a clear picture of the viability of solar pumping.

The contents of this book are based on a thorough, highly professional effort to ascertain the utility of solar powered pumping. In 1978 the United Nations Development Programme initiated a project, executed by the World Bank, for the "Testing and Demonstration of Small Scale Solar Powered Pumping Systems". The project was designed to assemble reliable technical and economic data from which to form a considered view of the viability of solar pumping systems. The project team examined the state of existing technology, carried out laboratory tests and field trials, and analysed the results over a period of 4 years. During the course of the project, and in major part as the direct result of it, small solar pumps have been developed to the stage where the best can meet all the technical and user prerequisites which are necessary to ensure satisfactory implementation of solar pumping on a wide scale.

The UNDP/World Bank project culminated in the production of a "Handbook on Solar Water Pumping", published by the World Bank in 1984. The Handbook was prepared by Intermediate Technology Power Limited, in association with Sir William Halcrow & Partners, the consultants for the UNDP/World Bank project. The authors condensed the detailed results of the project to produce a Handbook which can be used by engineers and decision makers who are considering using, buying, developing or selling solar pumps.

The Handbook has now been updated and is published here in the form of a practical introduction to solar water pumping. It is an important and timely contribution which should help to take solar pumps out of the realm of specialist research and development, and into the fields and villages of users who will have full knowledge of what the technology can do for them.

Dennis H. Frost
Chief Executive
Intermediate Technology Development Group
Chairman, IT Power Ltd

## ACKNOWLEDGEMENTS

The authors wish to express their appreciation to the other members of the IT Power/Halcrow Project Team for work carried out on which this book is based. These are Dr. David Wright, Dr Michael Starr, Peter Fraenkel, Bernard McNelis, Anthony Derrick, Chris Kropascy and Mike Aylward. The Project benefited from the constructive relations between the Project team and staff at the World Bank - Richard S. Dosik, New Energy Sources Adviser, Dr. Anwer Malik, Renewable Energy Sources Adviser, Dr. Norman Brown, Renewable Energy Specialist and Dr. Essam Mitwally, UNDP Project Manager.

JPK, WBG

January 1985.

# CONTENTS

This book has been written by authors from the Intermediate Technology Power Limited/Sir William Halcrow & Partners team which carried out the UNDP/World Bank Global Solar Pumping Project, between 1979 and 1983. The book presents a thorough and up-to-date review of solar pumping technology and its economics, based on real field experience combined with laboratory testing.

Since this work was completed IT Power and Halcrow have established a new joint company: Global Renewable Energy Services Limited, which operates the unique photovoltaic test facility originally established for the UNDP/World Bank Solar Pumping Project. GRES provides training courses on both practical and theoretical aspects of photovoltaic systems and offers testing and system development services to manufacturers.

# 1.  IS SOLAR PUMPING FOR YOU?

## 1.1  Purpose of this Handbook

Water pumping, which of course requires energy, is a basic need for a large proportion of the world's rural population. Since the majority of this rural population live in the sunny tropics or sub-tropics, to use the sun's energy is an attractive way of providing these vital energy needs. Traditionally, water is provided by hand or with the assistance of animals, while the principal source of mechanised power for rural areas of the world is the internal combustion engine. Recently there has been a revival of interest in windpumps as well as a growing interest in the new technology of solar powered water pumps.

There are two methods by which solar energy can be converted to the mechanical energy required for pumping. These are either (a) direct conversion of solar radiation to electricity using photovoltaic (PV) cells and then conversion of electrical energy to mechanical energy using a motor/pump unit; or (b) the conversion of solar energy to heat, which can then be used to drive a heat engine. The latter approach has received widespread attention in research institutions but so far no such systems have proved reliable.  Consequently, at present (1985), the most suitable approach to solar water pumping is to use photovoltaic (PV) powered pumps.

The technology of solar PV water pumping has been advancing steadily in recent years. From 1979 to 1983 the World Bank executed a UNDP funded project entitled "Small Scale Solar Powered Pumping Systems". The final report on this work provides an in-depth technical and economic analysis of the subject*. This handbook has been prepared by the Project consultants and summarises their experience and findings. Its purpose is to help the potential user to identify situations in which solar pumping should be considered, and also to show how the technical and economic details of such applications can be evaluated.

In this first chapter the basics of solar water pumping are reviewed briefly. An indication is given of the applications where solar pumps are likely to be viable so that the reader can rapidly establish if solar pumps are a feasible option for a particular situation, without going into the detailed analysis of the subsequent chapters.

-------------------------------------------------------------------------------

* "Small Scale Solar-Powered Pumping Systems: The Technology, its Economics and Advancement" Sir William Halcrow and Partners in association with Intermediate Technology Power Ltd (1983). Published by The World Bank, 1818 H Street, NW, Washington DC 20433, USA.

## 1.2  Energy for Water Pumping

The starting point for any assessment of water pumping is the relationship between energy and water requirements.  The pumping (or hydraulic) energy required to deliver a volume of water is given by the formula

$$E = \rho g V h$$

Where:

E is the required hydraulic energy in Joules*
V is the required volume of water in cubic metres ($m^3$)
h is the total head in metres (m)
$\rho$  is the density of water (1000 kg/$m^3$)
g is the gravitational acceleration (9.81 m/$s^2$)

With V in cubic metres and h in metres the pumping energy is

$$E = \frac{9.81 \ Vh}{1000} \qquad MJ*$$

For example:  To lift 60 $m^3$ through a head of 10 metres requires (9.81 x 60 x 10 ÷ 1000 ) = 5.89 MJ (1.64 kWh) of hydraulic energy.

Figure 1 illustrates the energy flows in a pumping system. The input energy for the pumping system undergoes several conversions before it is made available as useful hydraulic energy. Each conversion has an associated loss of energy which means that the input energy requirements for pumping are generally far greater than the useful hydraulic energy output. For example, if the power source (prime mover) is a diesel engine, the input energy will be in the form of diesel fuel: the energy content of 1 litre of diesel fuel is approximately equivalent to 38 MJ (10.5 kWh). If the fuel is converted to mechanical energy with an efficiency of 15% then 1 litre of fuel will produce 38 x 0.15 = 5.7 MJ (1.6 kWh) of mechanical energy. A pump may then convert the mechanical energy to hydraulic energy with an efficiency of say 60%, giving a useful hydraulic energy of 5.7 x 0.6 = 3.42 MJ (0.95 kWh) for 1 litre of fuel. In a similar way losses occur when solar radiation, muscle power or wind energy is converted to hydraulic energy.

In addition to energy conversion losses, a proportion of the pumped water may be lost in the process of delivering the water to its point of use. This will have a direct effect on the energy required for pumping and since, as will be shown in later chapters, the input energy requirements have a large influence on water costs, the efficiencies with which energy is converted and with which water is distributed, are of major importance.

-------------------------------------------------------------------------

* The Joule is the International System (S.I) unit of energy. It is best expressed in millions, as MegaJoules (MJ) because this is a more practical unit. The conversion rate to the more familiar kWh is 3.6 MJ = 1 kWh.

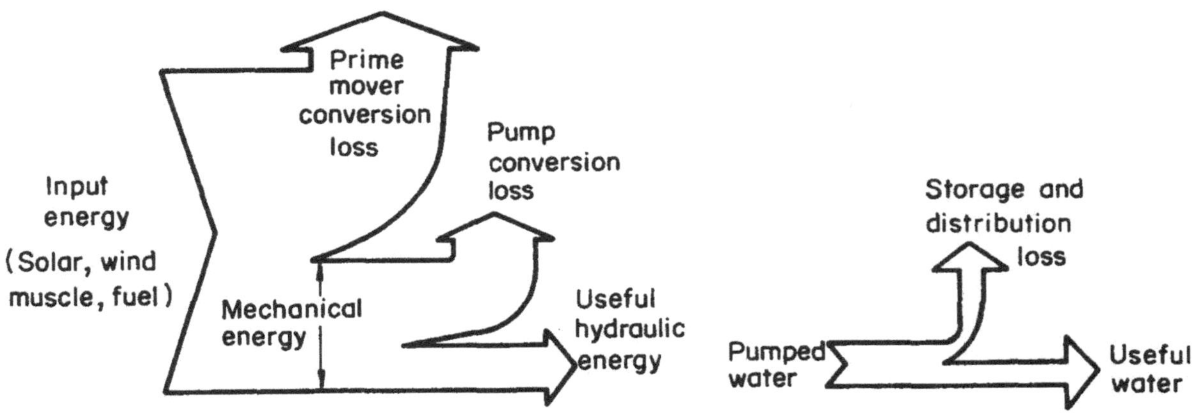


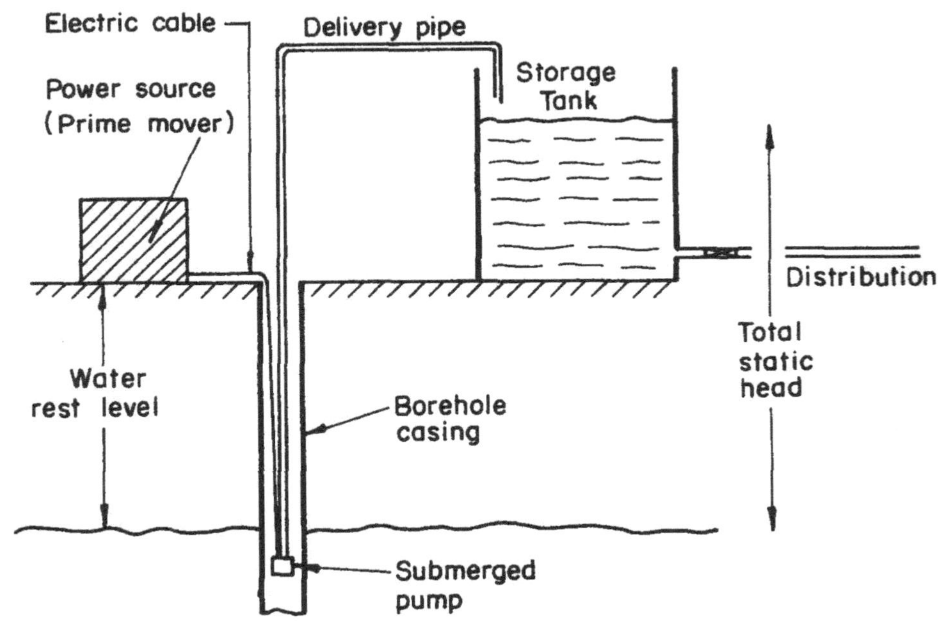


Figure 1. Schematic of a pumping system showing energy flows.

3

The power (P) required to lift a given quantity of water depends on the length of time that the pump is used. Power is the rate of energy supply, so the formula for hydraulic power is simply obtained from the formula for energy by replacing volume with flow rate (Q), in cubic metres per second.

$$P = \rho g Q h \qquad \text{Watts}$$

If the flow rate (Q) is in litres per second then the hydraulic power is:

$$P = 9.81\, Qh \quad \text{Watts}$$

For example, the average hydraulic power required to lift 60 m$^3$ of water through a 5m head in a period of 8 hours, (i.e. an average flow rate of 2.08 litres per second) would be 9.81 x 2.08 x 5 = 102 Watts. With a typical pump efficiency of 60%, the mechanical power required would be 102 ÷ 0.6 = 170 Watts.

_Energy (E)_ is the more important characteristic of water pumping since it is energy that has to be paid for in the form of diesel fuel, human labour, animal feedstock, or solar pump size. The equivalent power requirement only determines how quickly the required quantity of water is delivered and the rate at which the energy is used.

_The head (h)_ has a proportional effect on the energy and power requirements with the result that it is cheaper to pump water through lower heads. It consists of two parts: the static head, or height through which the water must be lifted, and the dynamic head which is the pressure increase, caused by friction through the pipework, expressed as an equivalent height of water. The static head can be easily determined by measurement and there are formulae for calculating the dynamic head. The latter depends on flow rate, pipe sizes and pipe materials. The smaller the pipes and greater the flow rate, the higher the pressure required to force the water through the pipes.


## 1.3  The Solar Energy Resource

The average value of the solar irradiance just outside the earth's atmosphere is approximately 1353 W/m$^2$. As solar radiation travels through the atmosphere it is attenuated and consequently the total power falling on a horizontal surface (known as the global irradiance) reaches a maximum of about 1000 W/m$^2$ at sea level. This is made up of two components, the radiation in the direct beam from the sun and diffuse radiation from the sky (radiation that has been scattered by the atmosphere).

Global irradiance varies throughout the course of the day because the path length of the solar radiation through the atmosphere changes. For the same reason there are variations with season and latitude and the total solar energy received in a day (known as the solar insolation or solar irradiation) can vary from an average of 2 MJ/m$^2$ (0.55 kWh/m$^2$) per day in the northern winter to an average of 20 MJ/m$^2$ (5.55

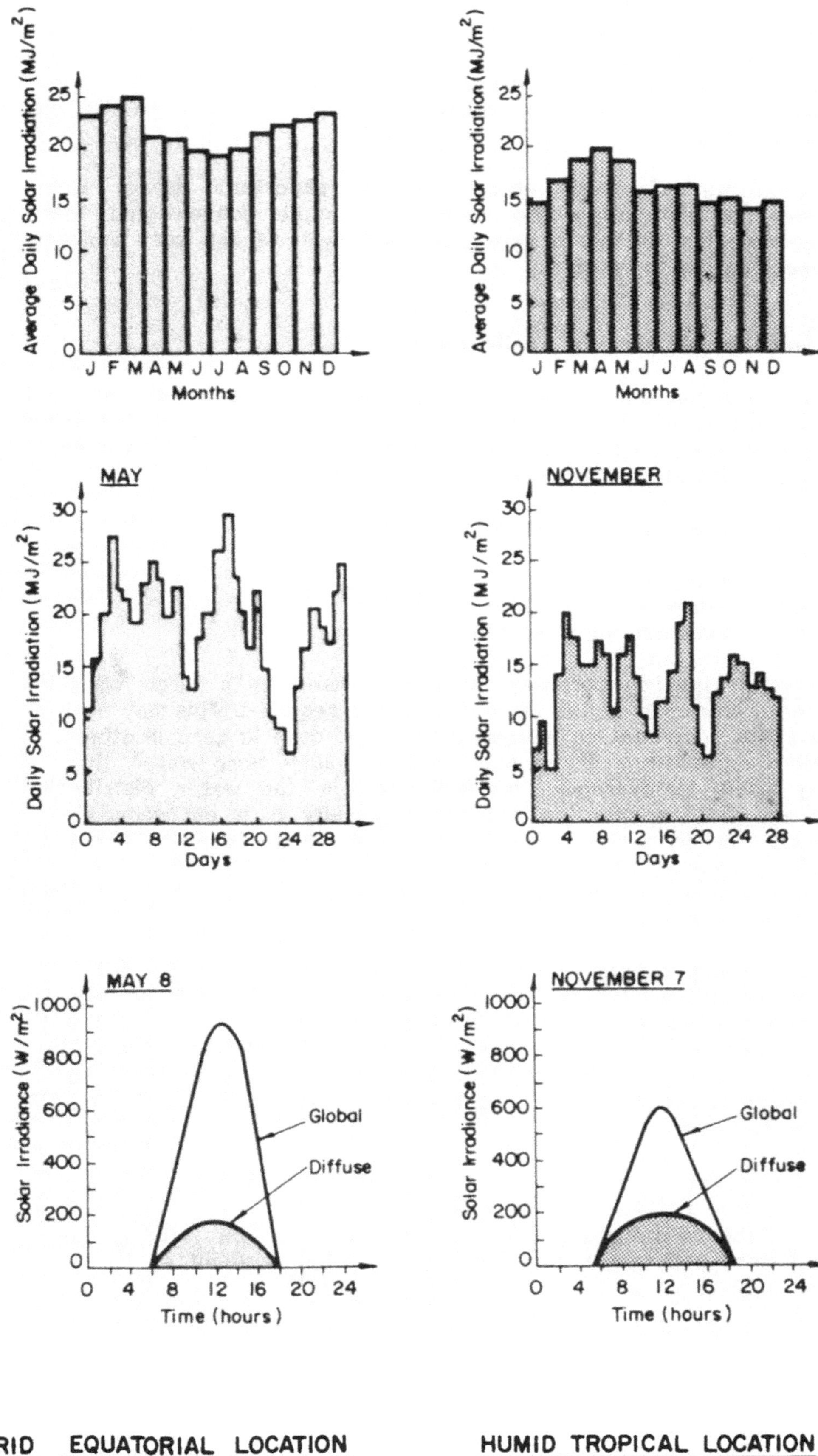



Figure 2. Typical hourly, daily and monthly variations in solar energy availability.

kWh/m$^2$) per day in the tropical regions of the world. On a clear day diffuse energy may amount to only 15-20% of the global irradiance whereas on a cloudy day it will be 100%. Figure 2 illustrates typical variations in solar energy throughout the day, from day to day and from month to month.

The variability of solar energy is an important aspect which influences system design and economics. Unlike conventional fossil fuel technologies the performance of solar systems can vary markedly from one location to another.


## 1.4 Typical Water Pumping Applications

It will be shown in the following chapters that solar pumps are most cost effective for applications with low power requirements, of the order of a few hundred watts. Fortunately this constraint is well matched to the majority of water supply needs in developing countries.

The three main applications for solar pumps are:

- o    irrigation
- o    village water supplies
- o    livestock water supplies

<u>Water for irrigation</u> purposes is characterised by a large variation from month to month in the amount of water required. This may peak at around 100 m$^3$/day/hectare in some months and drop to zero in others. In irrigation applications it is necessary to pump more water than is actually used to overcome inefficiencies in the water distribution system and field application methods. Generally it is not economic to lift water for irrigation through very high heads because increasing the lift increases the cost, and the cost of supplying water for irrigation should not be more than the value of the additional crop that can be grown.

Most developing country farms have areas in the range 0.5 to 2 hectares, which with typical water requirements of 20 to 80 m$^3$/day per hectare and pumping heads of 5 to 10 metres, give daily hydraulic energy requirements in the range of 1 to 8 MJ (0.28 to 2.2 kWh) per hectare. If the water is provided over an eight hour pumping period the average hydraulic power requirement is in the range 35 to 280 watts per hectare.

<u>Water for rural water supplies</u> (livestock or villages) is characterised by a more constant month by month demand. It is critical to have water available on demand so systems generally include a   storage tank. A widely accepted target for water consumption is about 40 litres per day per capita. At a figure of 40 litres per head, the daily volume which would have to be supplied to a village of 500 people would be 20 m$^3$. Pumping heads in water supply systems are typically greater than those in irrigation systems. Hence in a typical example with a head of 20 metres, the daily hydraulic energy requirement is about 4 MJ (1.1 kWh) and with a pumping period of 8 hours per day, the average hydraulic power requirement is about 140 Watts.

Similar sized systems are required for livestock water supplies. For example, cattle water consumption is also about 40 litre per head per day which, with a herd size of 500 cattle gives energy and power requirements close to those for village water supplies.

The value of water for domestic use is considerably higher than water for irrigation. For the relatively small quantities involved, people are prepared to pay high prices per cubic metre of water. Because of the more even month by month water demand and the higher value of the water, solar pumps for water supplies can be competitive at higher lifts than those for irrigation.

### 1.5  Overview of Solar Pump Viability

Since a solar pump consists essentially of an electrical source (the P.V array) powering a motor/pump unit, it is technically possible to assemble a solar pump for all applications where an electric pump can be used. Thus solar pump viability centres on the question of costs:

'How does the cost of water provided by a solar pump compare with the cost of water provided by alternative methods?'

In order to make an appraisal of the feasibility of using a solar pump, the following properties must be estimated:

(a)  Water demand factors

o    Average annual (crop, village or livestock) water requirement in cubic metres per day (V). This is obtained by dividing the annual water requirement by 365 days.*

o    Total lift in metres (h)

o    Peak monthly crop water requirements in cubic metres per day obtained by dividing the maximum monthly crop water requirement by the number of days in the month. This is only required for irrigation systems so that ratio of peak requirements to average requirements can be determined.

---

* Note that irrigation water requirements are normally specified in mm – 1mm is equivalent to 10 $m^3$ per hectare. Also the irrigation season is generally only part of the year, so the average irrigation water requirement will be somewhat lower than the actual daily use. The average is simply a measure of the annual requirement and together with the peak water requirement gives an estimate of the variability of water demand. Note that it is the water requirement of the crop that must be estimated – the volume of water pumped will be considerably higher than this because of distribution losses.

(b)  Location dependent factors

> o    Minimum average monthly wind speed in m/s (u) (to
> o    investigate whether a windpump might be more economic)
>
> o    Minimum average monthly solar irradiation in
> o    MJ/m$^2$/day (H) since for sunnier locations the
> o    economics of solar pumping are more attractive.

(If these are not immediately known, methods for obtaining them are described or referenced in later chapters).


From these properties make the following simple calculations:

(i)  Multiply the average annual water requirement (V) in m$^3$ per day by the total lift (h) in metres to give the average **energy equivalent** (Vh) in m$^4$ per day. The energy equivalent is a measure of the end-use water requirement and is not a true measure of the hydraulic energy requirement because of distribution and field application losses. However it is more important than the hydraulic energy because it is the cost of water delivered to the user that should be used to judge alternative options.

(ii) For irrigation systems, divide the peak monthly water requirement in m$^3$ per day by the average monthly water requirement (V) in m$^3$ per day to give the peak demand factor (PDF). The PDF has a very significant effect on solar pump economics since the photovoltaic array must be sized to provide the peak requirements and as the peak is significantly higher than the average, the solar pump is under-utilised for much of the year.

As a general aproximation it can be shown that solar pumping systems for irrigation are beginning to become cost competitive compared to diesel pumps in situations where the <u>peak</u> daily energy equivalent is less than about 150 m$^4$ (for example 30 m$^3$/day through a head of 5 m ) <u>and</u> where the minimum monthly average solar irradiation is greater than about 15 MJ/m$^2$ per day. For windy locations where the minimum monthly average wind speed is greater than 3 m/s a windpump would be a cheaper option.

Similarily for rural water supply applications solar pumping systems are becoming cost competitive compared to diesel pumps where the <u>average</u> daily energy equivalent is less than about 250 m$^4$ (for example 25 m$^3$/ day through a head of 10 m) <u>and</u> where the monthly average solar irradiation is greater than 10 MJ/m$^2$ per day. Windpumps are generally cost competitive at locations with minimum monthly average wind speeds greater than 2.5 m/s.

There are, of course, important factors not captured in a straight cost comparison and these should be taken into account when one is comparing pumping methods and making choices. For example when considering a diesel pump, the reliability of fuel supplies and availability of spare parts should be taken into account. An unreliable fuel supply may make a diesel pump inoperable for part of the year with no guarantee that such a cost comparison may not create an open and shut case for the choice of a pumping system and in certain circumstances (e.g. remote, inaccessable locations) the greater reliability of a particular pumping system may offset its higher cost.

To help the reader make an initial appraisal of the feasibility of using a solar pump, the decision chart in Figure 3 has been prepared. It refers only to the major mechanised options for water lifting, i.e. wind, solar and diesel, and is based on the unit water costs shown in Figures 27 and 28. It cannot be used to compare solar pumps with hand or animal powered pumps.

<u>Use of the Decision Chart for Solar Pump Viability Appraisal</u>

Trace a path from the starting point of the chart for the particular values of energy equivalent (Vh), peak demand factor (PDF), solar irradiation (H) and wind speed (u). Figure 3a is used for irrigation pumps and Figure 3b for rural water supplies. Choices are represented by diamond shaped boxes. The assessment is given when a rectangular box is reached.

<u>Consider using a windpump.</u> For relatively windy locations windpumps can provide a cheap, reliable water supply. However wind speed data are often unreliable and the cost of water from a windpump is very sensitive to monthly mean wind speed; hence it is important to ensure that the wind speed data are accurate. The wind regime is also very site sensitive and sheltered locations (in woods or valleys) may prove unsuitable while nearby hill tops provide ideal conditions. Windpumps are a time proven technology requiring little maintenance and having long operating lifetimes.

<u>Consider using a diesel pump.</u>  At present, where water requirements are high or the minimum monthly solar irradiation is less than $15\,MJ/m^2$ for irrigation or $10\,MJ/m^2$ for rural water supply, diesel pumps can provide cheaper water. However, as noted above, the reliability of fuel supplies and availability of spare parts should be taken into account when considering a diesel pump.

<u>Consider using diesel for short term, reconsider solar pumping as long term option.</u> This decision indicates that solar pumps are not the cheapest option at present, but if anticipated reductions in PV module costs are achieved, then solar pumps will become more economically viable.

<u>Consider using solar pump if water demand is low, otherwise diesel pump.</u> This indicates diesel powered water pumps are probably the cheaper option. A more thorough analysis should be made if water requirements are small, or well matched to the available solar energy, because the smallest diesel engines (rated at 2.5 kW) are considerably oversized for an energy equivalent less than $50\,m^4$.

<u>Consider using a solar pump.</u> This decision will generally be reached for low energy equivalent (Vh) and sunny locations. It indicates that solar pumps should be considered by making a further analysis as detailed in Chapter 3.

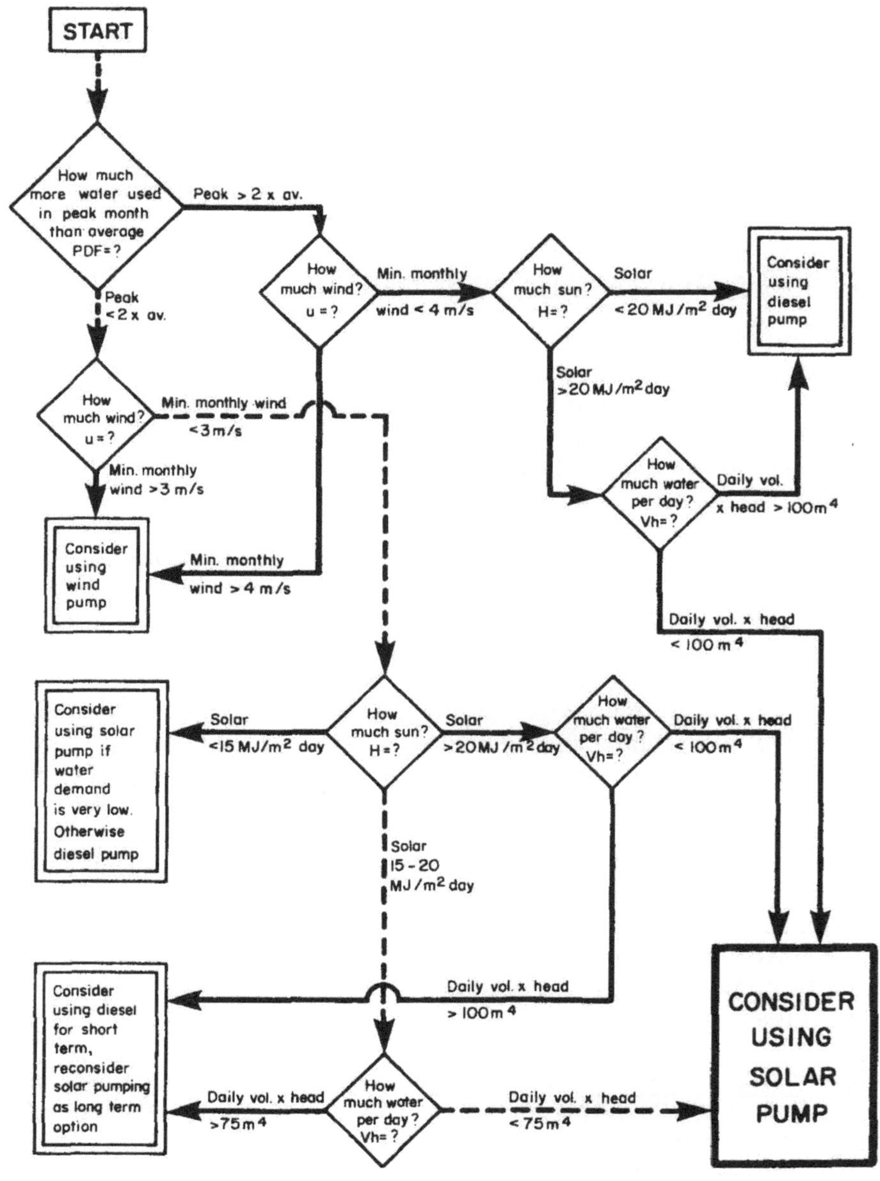


Figure 3a. Decision chart for an appraisal of solar pumps for irrigation

10

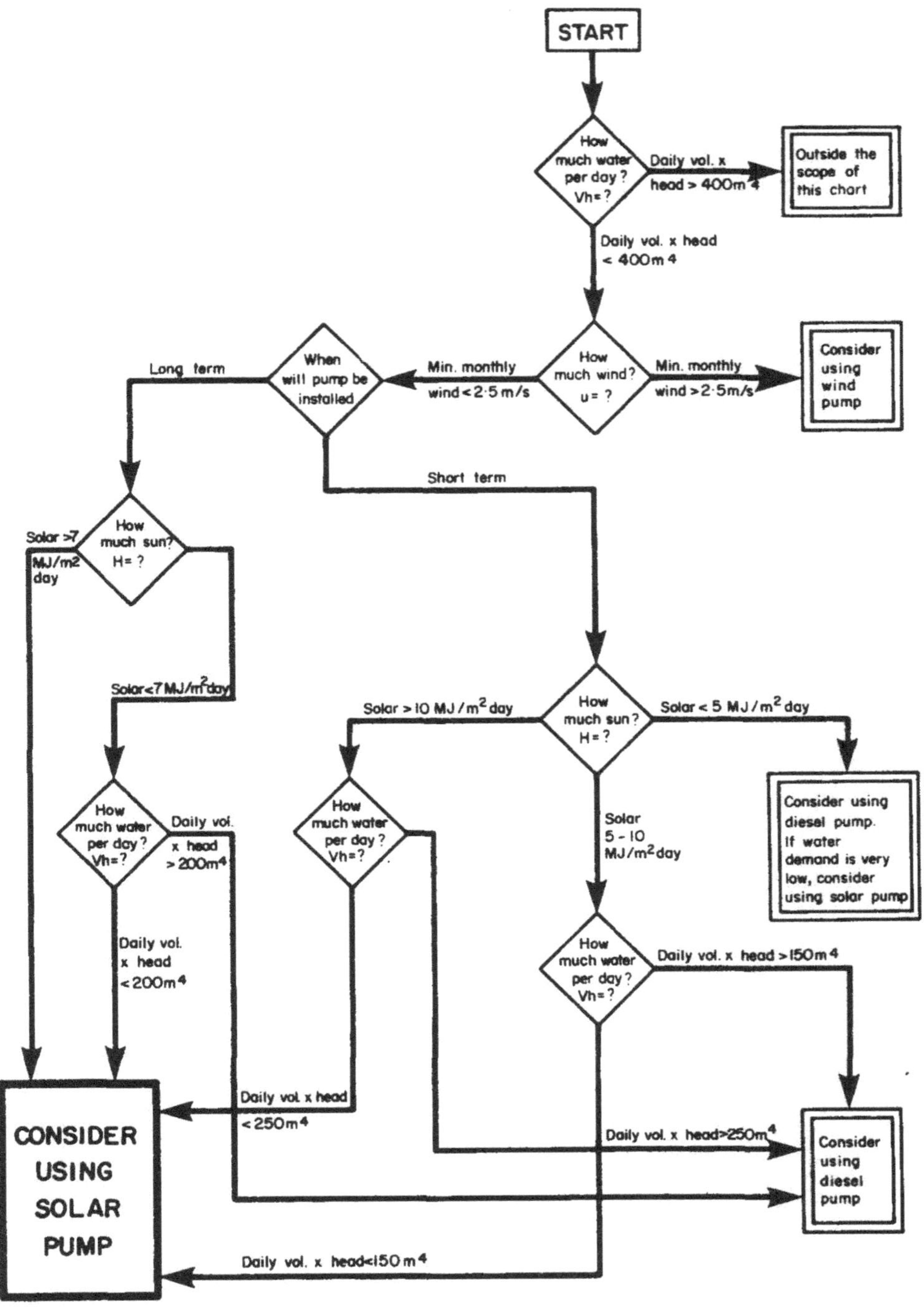


Figure 3b.    Decision chart for an appraisal of solar pumps for rural water supply

<u>Examples to demonstrate the use of the Decision Chart.</u>

Consider a 1 hectare irrigation system with a water lift (h) of 2 metres; an annual crop water requirement of 1277.5 mm and a peak monthly crop water requirement of 6 mm per day. The minimum average monthly solar insolation (H) and wind speed (u) are 17 MJ/m$^2$ (4.7 kWh/m$^2$) and 2.0 m/s respectively. The stages below illustrate how an initial appraisal of solar pump viability can be obtained.

(i) for a 1 hectare plot the annual crop water requirement is 12775 m$^3$ which is equivalent to an average of 35 m$^3$ per day. The peak water requirement is 60 m$^3$per day.

(ii) therefore the Peak Demand Factor is 60 ÷ 35 = 1.71

(iii) the average annual energy equivalent is (35 x 2) = 70 m$^4$per day;

(iv) a path, indicated by the dotted line in Figure 3a, may now be traced from the START for the above values of H, u, Vh and PDF. The first choice is for the peak demand factor. In this case the peak demand factor (PDF) is 1.71, so the 'Peak < 2 av' path is followed. The second choice is for wind speed. In this case wind speed is less than 3 m/s so wind pumping is not selected. The third choice is for solar irradiation. For the 15-20 MJ/m$^2$ band, a path is followed to a 'How much water' box. With an energy equivalent less than 75 m$^4$ an assessment is reached 'Consider using solar pump'. This indicates that for this particular application and location, the option of using solar pumps should be seriously studied.

As a further example consider a village water supply system with a requirement of 10 m$^3$ per day in a location with a minimum monthly solar insolation of 15 MJ/m$^2$ (4.2 kWh/m$^2$)per day and a minimum monthly average wind speed of 2 m/s. The head is 20 m:

(i) the average energy equivalent is (10 x 20) = 200 m$^4$ per day

(ii) a path is traced on Figure 3b. The first choice is for energy equivalent. For Vh less than 400 m$^4$ the chart leads on to wind speed choice. Since, for the example, the wind speed is less than 2.5 m/s, wind pumps are not viable and the choice is between solar and diesel pumps. Both in the short term and the long term the choices lead on to the assessment 'Consider using solar pump'.

The procedures for a more accurate assessment of solar pump viability and the state of the technology at present are described in the following chapters.

## 2.    THE TECHNOLOGY

### 2.1  Photovoltaic Pumping Systems

The main components of a PV pumping system are illustrated schem-
atically in Figure 4. A solar PV pumping system can be divided
conceptually into three parts:

o the PV array which converts solar energy to d.c electricity

o the motor and pump subsystem comprising the components which
convert the electrical output of the PV array into hydraulic
power

o the storage and distribution system which delivers the water to
its point of use

The capital cost of PV arrays is directly proportional to the electrical output of the array and the PV array is at present a large
proportion of the overall system cost. Consequently the cost of a
solar pump is almost directly proportional to the hydraulic output and
there are only economies of scale for small systems.

The efficiency of the subsystem determines how large the PV array
needs to be for a particular hydraulic duty and hence has a large
influence on the overall solar pump cost - a more efficient subsystem
will require a smaller PV array and the solar pump will cost less.

It is common for little thought to be given to the design of the
delivery side of a pumping system. In many cases, especially in
irrigation systems, water is wasted and far more water than is
actually required for the crop must be provided by the pump. For a
conventional diesel pump this means pumping for a longer period which
means more fuel will be consumed, maintenance costs will increase and
the diesel engine lifetime will be shortened. This increases the
system running costs, but has no effect on the capital cost of the
pump. The effect of a wasteful distribution system is far more significant for a solar pump. Since the capital cost of a solar pump is
almost proportional to the quantity of water pumped, it is particul-
arily important to ensure that the water is distributed efficiently and
to take into account the distribution system when assessing the
overall system costs.

With conventional gasoline or diesel fuelled pumps, storage is not
important because energy is stored in the fuel itself and the pump can
be started when there is a demand. With a solar pump, energy is not
available on demand and the day by day variations in solar irradiation
mean that for some locations, with successive cloudy days, it may be
prudent to consider storage of a surplus of water pumped on sunny days
for use on cloudy days.

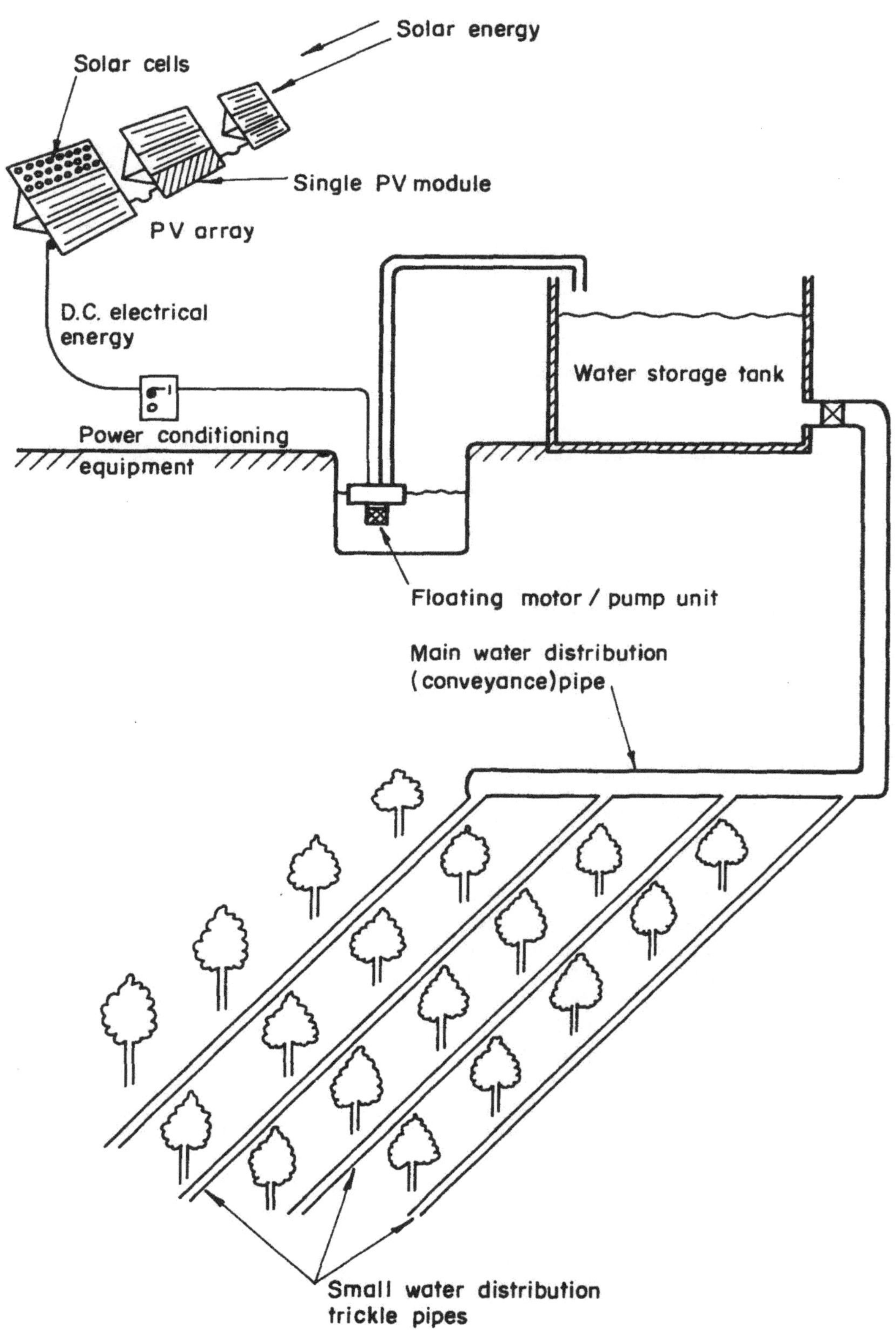


Figure 4.   Main components of a PV pumping system showing a piped distribution for irrigation.

## 2.2  Photovoltaic Arrays

### 2.2.1  Solar Cell Technology

The main type of photovoltaic cell used commercially for solar water
pumping at present is the monocrystalline silicon cell although other
types of silicon cell, such as amorphous silicon and polycrystalline
silicon cells can be expected to play an increasing role in the market
in the future. A number of semi-conductors are being studied as
alternatives to silicon, such as cadmium sulphide and gallium arsen-
ide, but these are not yet commerically available as photovoltaic
modules. Pure silicon is an electrical insulator, but if traces of
certain impurities are added to it, it will conduct electricity. To
make mono-crystalline solar cells, large silicon crystals 'doped' with
impurities are grown and sliced into thin wafers usually 100mm in
diameter. A second chemical impurity is diffused into one surface of
the wafer. When solar radiation strikes this surface an electrical
potential difference is generated. Metal contacts attached to each
side of the wafer allow this to create a current flow. Figure 5
illustrates the principles involved.

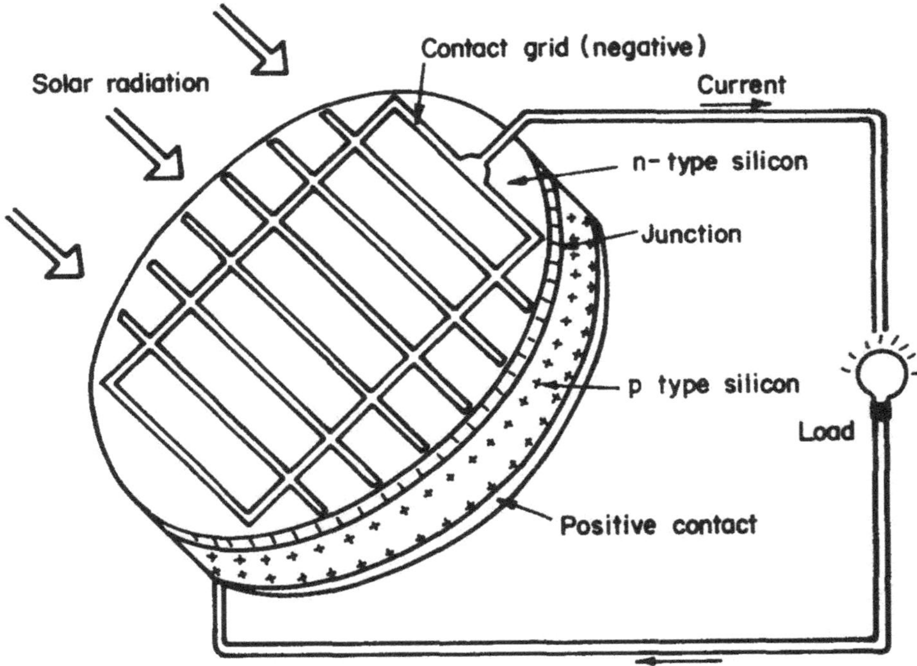


Figure 5.  Construction of a silicon photovoltaic cell.

Under an irradiance of 1000 W/m$^2$ (near to the maximum at sea level)  a
typical PV cell will deliver a maximum of about 25 mA of current for
each square cm of cell surface. If the cell is left open circuit
(i.e. unconnected to a load) then it will generate a voltage of 0.6
Volts. The efficiency of conversion from sunlight to electrical power
is typically 10 - 13%. Thus a single cell of diameter 100 mm will
generate nearly 1 watt under an irradiance of 1000 W/m$^2$.

Cells are electrically connected in series and parallel to give suitable voltages and currents for a particular application. A number of cells are encapsulated into a module, and these modules are the building bricks of a PV system. A typical module is sized about 1 m by 0.3 m by 50 mm thick and contains 36 cells, thereby producing an output of 30 – 35 Watts at 12 volts in bright sunshine.

Several modules are combined to form a PV array. (A typical example is shown in Figure 6). Galvanised steel, aluminium or chemically treated wood have been used to support the modules. The former two are likely to be the most durable in developing country environments. Arrays are usually fixed in position and mounted on concrete foundations. Some arrays can track (follow) the apparent motion of the sun and so intercept more energy but this can also increase cost and complexity. Portable arrays can be of use if the pump is required to operate at different locations (for example to supply a field from a canal at several different places).

Figure 6.    Photovoltaic arrays.

## 2.2.2  Performance

Photovoltaic modules are rated in peak Watts (Wp).  This is a reference value of the maximum power output from the module when operating at a cell temperature of 25 °C under a solar irradiance of 1000 W/m$^2$, and is a higher power output than is achieved on average in the field.
If the cell efficiency is 10%, a 1 kWp array would contain a cell area

of 10 m$^2$. Typically the <u>packing factor</u> (the ratio of cell area to array area) for circular cells is about 75% giving a gross area of 13.3 m$^2$ for a 1 kWp array. The packing factor can be increased by using square cells, but this involves cutting the cells. As cells become cheaper to produce, higher packing factors can be expected.

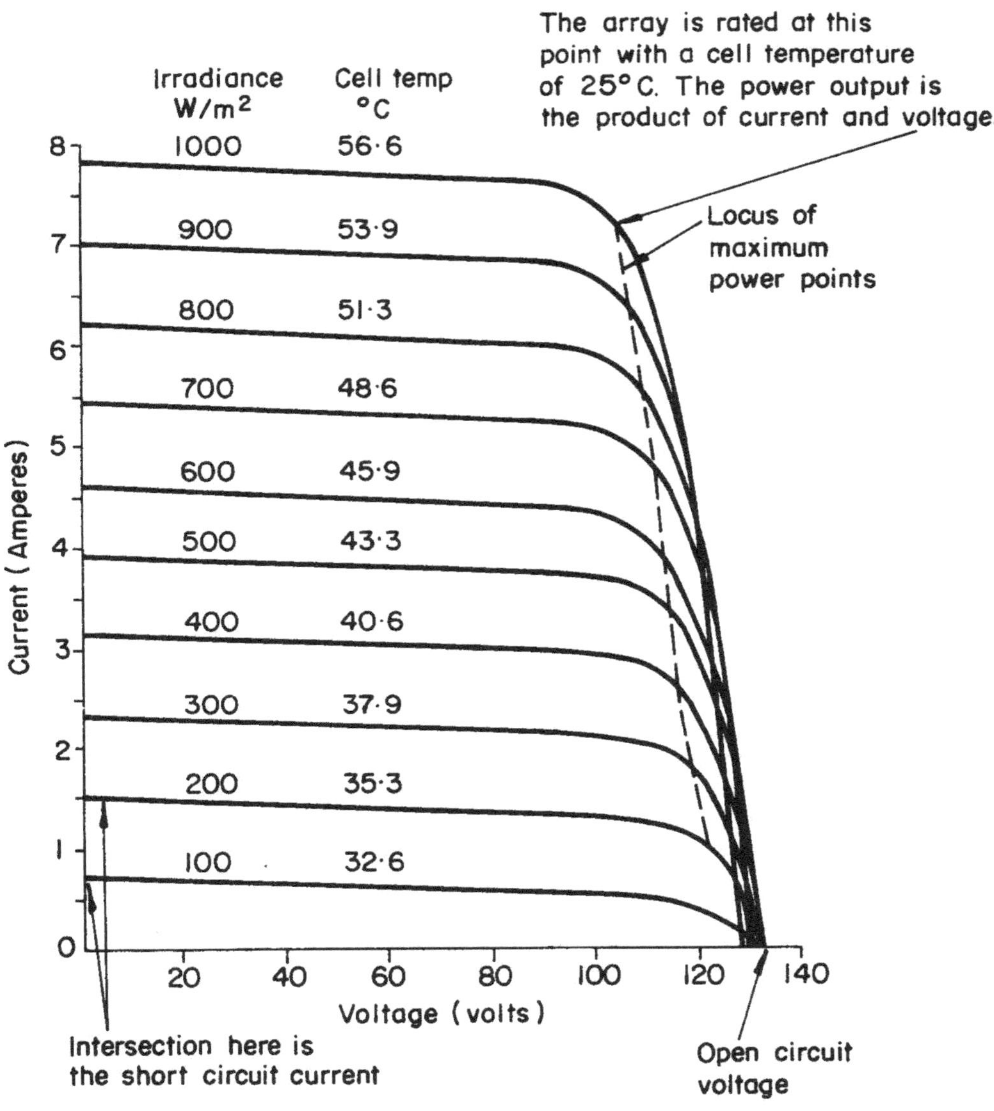


Figure 7.    A typical relationship between current and voltage for a PV array.

The efficiency (and hence power output) of the cell depends on the electrical load because of the relationship between current and voltage for the cells. A typical current/voltage (I/V) characteristic for an array is shown in Figure 7. It can be seen in this figure that the cells have a varying internal resistance depending on their operating point (i.e. the values of current and voltage are determined by the load). The power output of the array is the product of voltage and current; hence for a given solar irradiance it will vary with the operating point. There is a maximum power point which occurs at the knee of the I/V curve. Under short circuit or open circuit conditions the power output is zero while the maximum power output is typically about 0.7 times the product of open circuit voltage and short circuit current. The ratio of maximum output power to the product of open circuit voltage and short circuit current is known as the <u>fill factor</u>.

The efficiency of solar cells falls off with increasing operating temperature. Typically the efficiency will drop by 0.5% fractionally per degree centigrade increase in operating temperature. With the daytime ambient temperatures found in many developing countries, cell operating temperatures can be as high as $60\,^{\circ}C$, resulting typically in a 16% fractional reduction in the peak output from that at the reference operating temperature. Also since solar irradiance is generally less than 1000 $W/m^2$, the average power output from an array is always significantly less than the rated output. In a good 'solar' location, the solar irradiance averaged over the hours of daylight may be 500 $W/m^2$. Hence the average array output will generally be less than half of the rated output.


## 2.3  The Motor and Pump Subsystem

### 2.3.1  General

In many cases it is feasible to utilise off the shelf, mass produced motors and pumps. However, special pumps and motors have been developed by some manufacturers with an above average efficiency to minimise overall system costs. The operation of the subsystem is unlike conventional power conversion devices, because the power supply varies as the incident solar energy changes. This means that the sub-system must be designed to work efficiently over a range of voltage and current levels.

Because of these variations and the resulting changes in subsystem efficiency, it is useful to define two types of efficiency:

> (i) the <u>power efficiency</u> of the subsystem, which is the ratio of hydraulic output power to electrical input power, at any instant in time. This will have a peak value when the sub-system operates at its design conditions;

> (ii) the <u>energy (or daily) efficiency</u> of the subsystem, which is the ratio of hydraulic output energy to electrical input energy over a day. It is a time average of the power effic-iency and consequently depends on the daily variations in power efficiency and hence on the solar irradiance profile for the day.

The energy efficiency is the more important parameter because it determines the array size for a particular hydraulic duty and con-sequently to a large extent how much the solar pump costs.

In Figure 7 it was shown that the output from a PV array depends on the electrical load, and that there is a maximum power point for each solar irradiance level. To maximise the output from the array the operating point (i.e. voltage and current of the subsystem) needs to be as close as possible to the knee of the array current/voltage curve. If the electric motor is coupled directly to the array then the system will operate at the point where the current and voltage characteristic of the motor intersects with the I/V curve of the array for the prevailing level of solar irradiance. Since this is not always at the optimum point, electronic power conditioners can be used to ensure that the output power is held as close as possible to its optimum value. It is also important that the efficiency of the subsystem should not decrease significantly when operating away from its design conditions, because the input power will vary with the solar irradiance level throughout the day.

The subsystem requires a minimum power input to start working. In the case of positive displacement reciprocating pumps this is because the motor has to overcome the peak starting torque of the pump. In the case of centrifugal pumps, the pump will usually rotate at very low irradiance levels but water will not be lifted until a threshold power level is reached and this threshold increases with the required pumping head. This means that the stopping and starting threshold levels of solar irradiance are important characteristics of solar pumps. A typical starting threshold is about 300 $W/m^2$. This means that on overcast days, when the solar irradiance on the array does not exceed about 300 $W/m^2$, a solar pump may not operate at all.

### 2.3.2 Motor Technology

Solar water pumps that are currently available make use of the following motor technologies

- brushed type permanent magnet d.c motors
- brushless permanent magnet d.c motors
- a.c motors

The choice of a d.c motor is of course attractive because the array provides a d.c power supply. However for high power applications a.c motors in conjunction with d.c - a.c inverters can be used. The range of a.c motors available to the designer is far greater than that of d.c motors, and the prices are generally lower. However, for small systems the savings made by using a low cost a.c motor may be offset by the additional cost of an inverter.

Permanent magnet brushed and brushless d.c motors are shown in cross section in Figure 8. Also shown is a conventional wound field d.c motor for comparison. In the conventional motor both the magnetic field of the armature and the surrounding field coil are powered by the d.c supply. The field in the armature is cycled by use of a commutator, thus causing the armature to rotate.

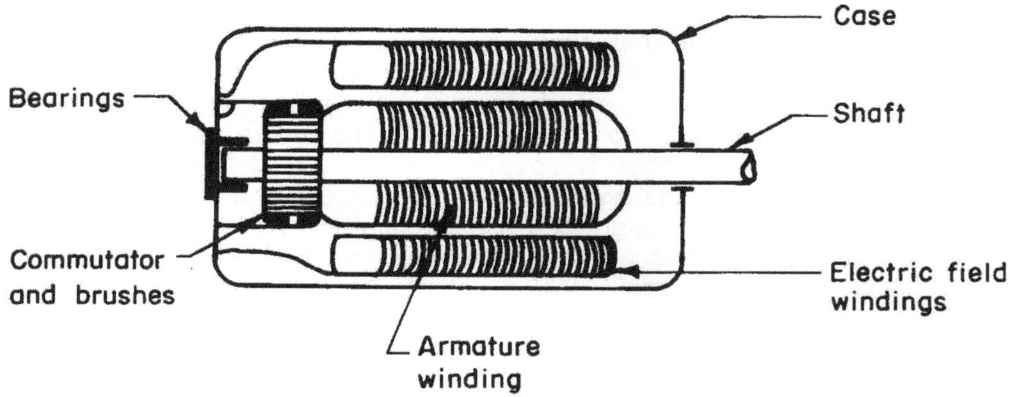




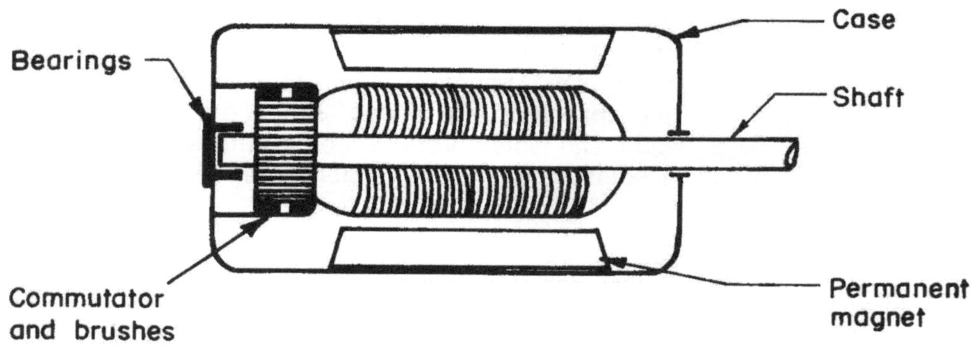




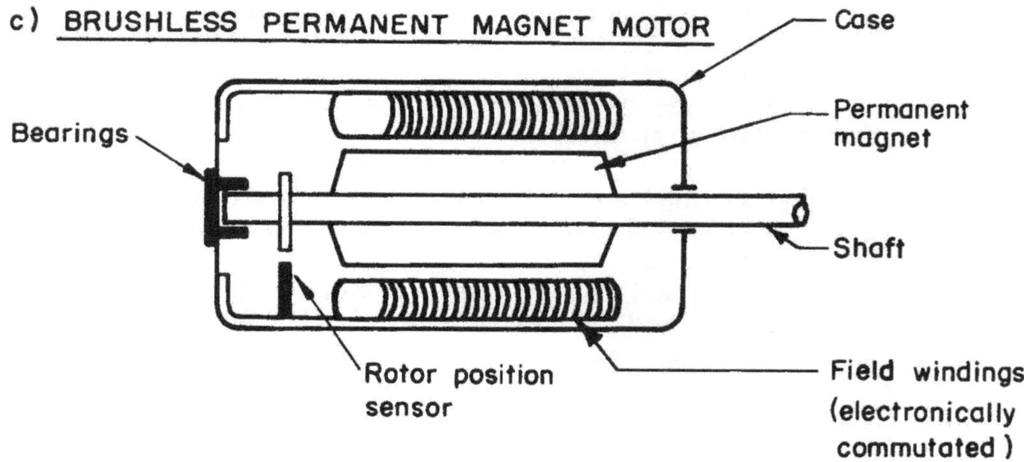


Figure 8.   Construction of d.c motors.

20

The permanent magnet brushed d.c. motor also achieves rotation by having a brushed commutator but the surrounding magnetic field is not induced electrically. This leads to higher efficiencies as no power is consumed in the field windings (and in turn lower PV array sizes may be used for the same hydraulic duty).

Brushed motors generally require new brushes at intervals in the order of 2000 - 4000 hours (equivalent to one to two years of continuous use with a solar pump) with obvious implications for maintenance costs. The brushless d.c. motor has a permanent magnet rotor and electronically switched field windings. Brushless d.c. motors in use with solar pumps are, at present, relatively new and early reliability problems with the electronics are being tackled by the industry. The long term potential for brushless d.c motors shows every likelyhood of being large.

The use of an a.c motor in a solar pump requires an inverter which introduces additional costs and some energy losses. Hence a.c motors have not been seriously suggested for low power (less than 250W) applications where the increased cost may be a significant proportion of the overall cost. A.c motors are generally less efficient than d.c motors but special improved efficiency models are now available for use in solar powered systems.

### 2.3.3. Pump Technology

In the design of a solar powered pumping system the pump itself is the most important component. Pumps can be divided into two categories centrifugal and positive displacement, and they have inherently different characteristics.

<u>Centrifugal pumps</u> (Figure 9) are designed for a fixed head and their water output increases with rotational speed. They have an optimum efficiency at a design head and a design rotational speed. At heads and flows away from the design point their efficiency decreases. However, they offer the possibility of achieving a close natural match with a PV array over a broad range of operating conditions. Centrifugal pumps are seldom used for suction lifts of more than 5 to 6 metres and are more reliably operated in submerged floating motor/pump sets. This is because they are not inherently self-priming, and can easily lose their prime at higher suction heads.

<u>Positive displacement pumps</u> (Figure 10) have a water output which is almost independent of head but directly proportional to speed. This means that the efficiency of a pump of fixed piston diameter increases with head and therefore for optimum efficiency different diameter pumps need to be used for different heads. At high heads the frictional forces become small relative to the hydrostatic forces and consequently at high heads positive displacement pumps can be more efficient than centrifugal pumps. At lower heads, below about 15m, the total hydrostatic forces are low in relation to the frictional forces and hence positive displacment pumps are less efficient and less likely to be used.

A major factor to consider when coupling a positive displacement pump

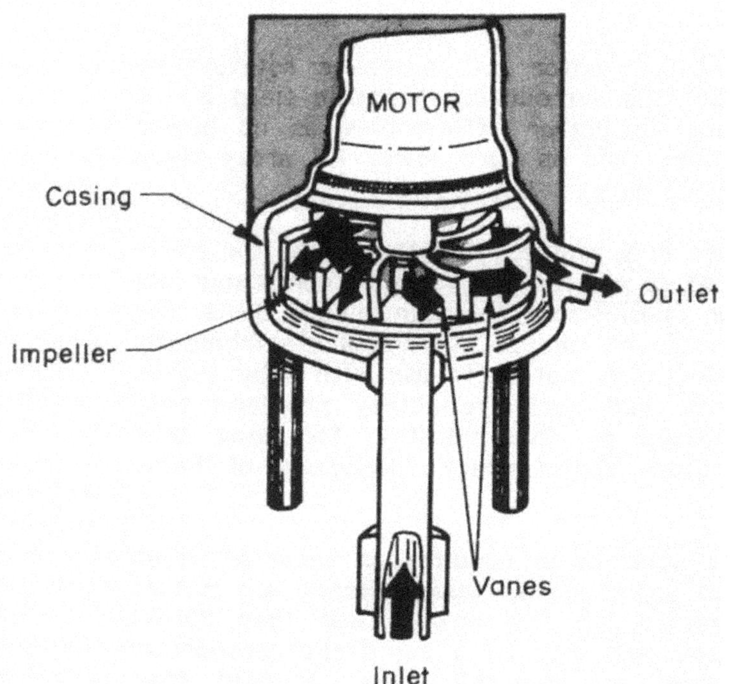


Figure 9. Principle of a centrifugal pump. Water is thrown out from the centre of the pump because of the centrifugal force created as the impeller rotates.

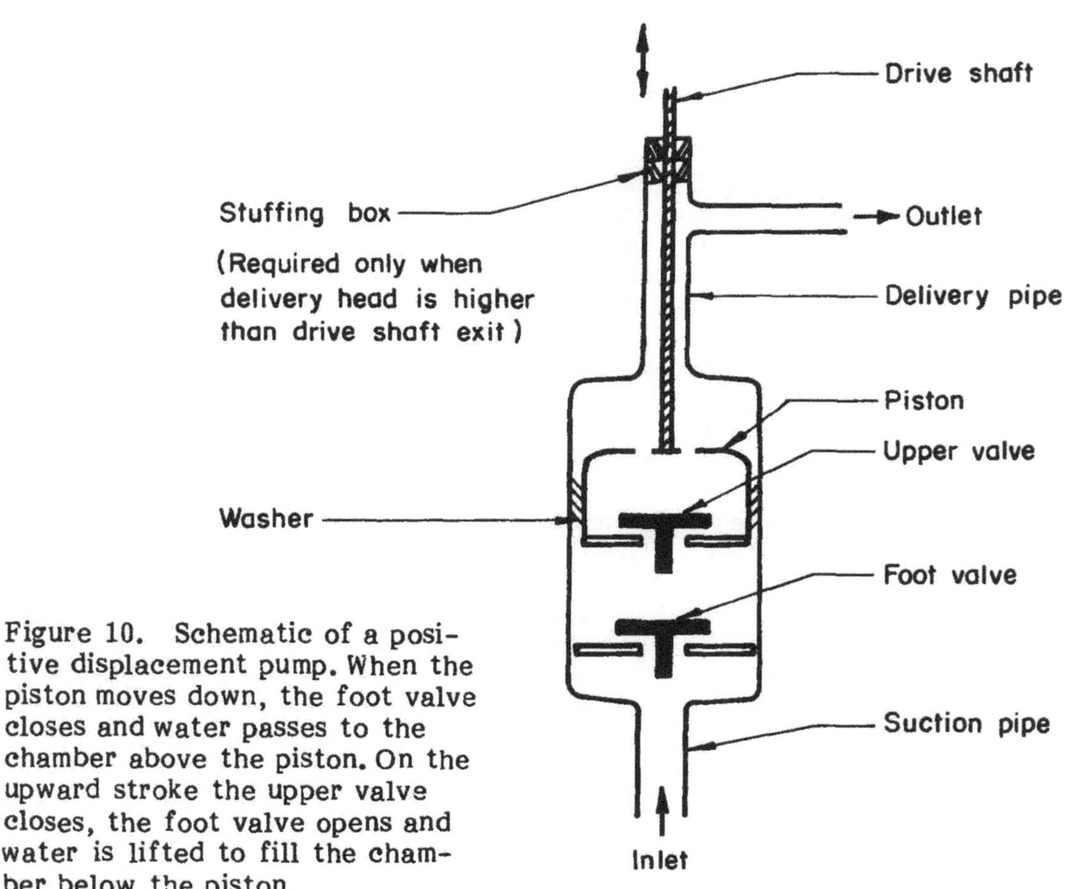


Figure 10. Schematic of a positive displacement pump. When the piston moves down, the foot valve closes and water passes to the chamber above the piston. On the upward stroke the upper valve closes, the foot valve opens and water is lifted to fill the chamber below the piston

to a PV array is the cyclic nature of the load on the motor. This causes cyclic variations in electrical impedance corresponding to the variations in torque. Electronic power conditioning is sometimes used to smooth out these effects by matching the current/voltage characteristics of the array with those of the motor. It is also important to match the motor operating characteristics to those of the pump by choosing appropriate gear ratios. Additional smoothing can be provided by the addition of a fly wheel.

### 2.3.4.  Power Conditioning

Power conditioning may be of several types:

- o   impedance matching devices
- o   d.c to a.c inverters
- o   batteries
- o   switches, protective cut outs etc.,

In almost all cases the use of power conditioning equipment implies a power loss (typically 5%), additional cost and an additional potential failure mode. Hence to justify their use the increased costs must be compensated for by the extra water output or in the case of switches and protective cut outs, better reliability.

<u>Impedance matching devices</u> are used:

> (a)  to produce high currents so that the motor pump will start in low solar irradiance conditions. (This is particularly important when using reciprocating pumps.)

> (b) to maximise the power available from the PV array.

Maximum power point trackers (MPPT's or maximum power controllers) which are "intelligent" devices, usually employing a micro processor, achieve these functions by sampling the power output of the array at frequent intervals (typically every 30 milli-seconds). They compare each new value of the array output power with the previous value, and if the power output has increased then the array voltage is stepped in the same direction as the last step, whereas if the power output has decreased then the array voltage is stepped in the opposite direction. The power consumption of maximum power controllers is typically between 4 and 7% of the array output.

An alternative method of impedance matching is to fix the PV array voltage. As can be seen in Figure 7 the array will then operate close to its maximum efficiency over a wide range of irradiance levels.

<u>D.c to a.c invertors</u> for use with PV arrays are currently undergoing considerable development and can be expected to become increasingly important. To maximise their benefit, the electronics involved should also provide some means of impedance matching such that a PV array can operate near to its maximum power point. The efficiencies of some commercially available inverters are claimed to be greater than 90%. Some inverters have poor part load efficiency and are therefore unsuitable for use with solar pumps.

23

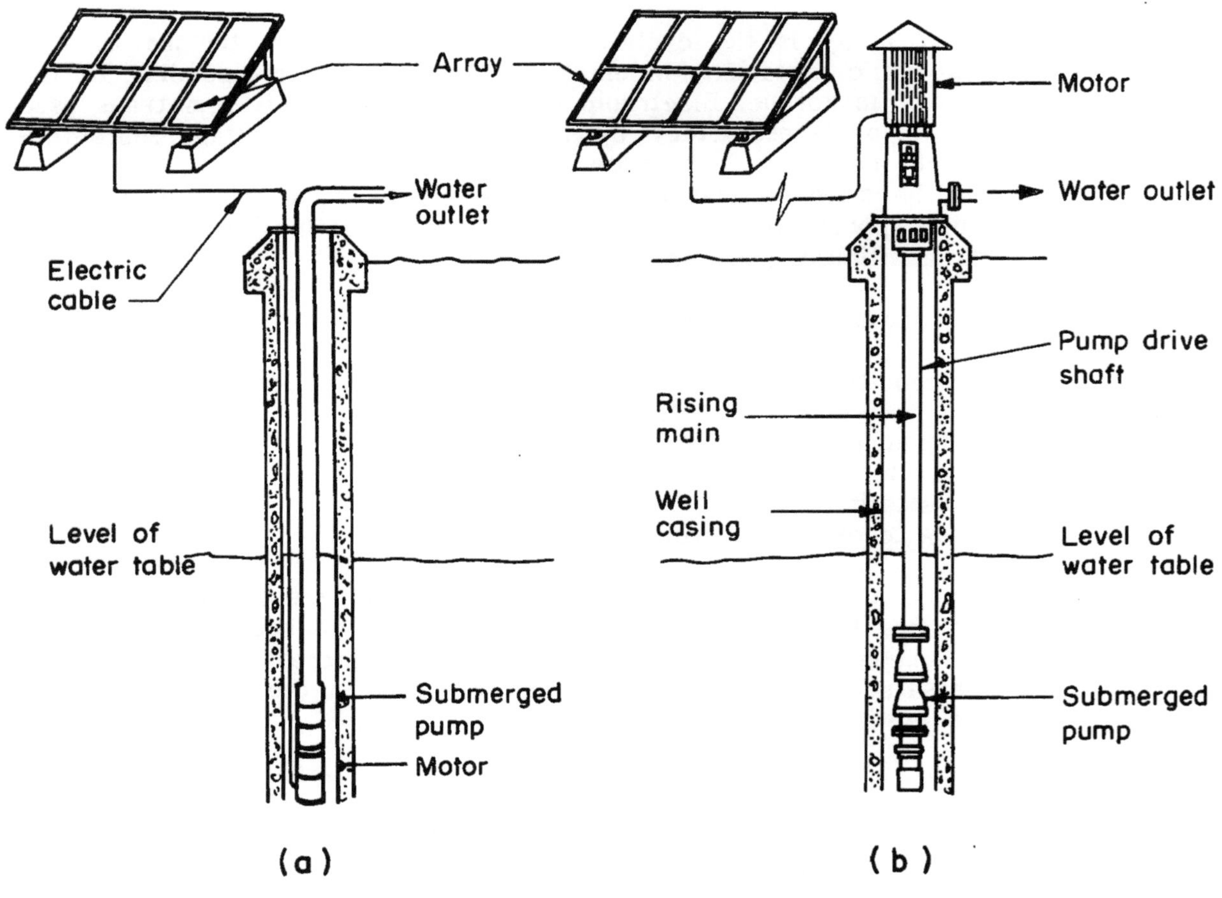


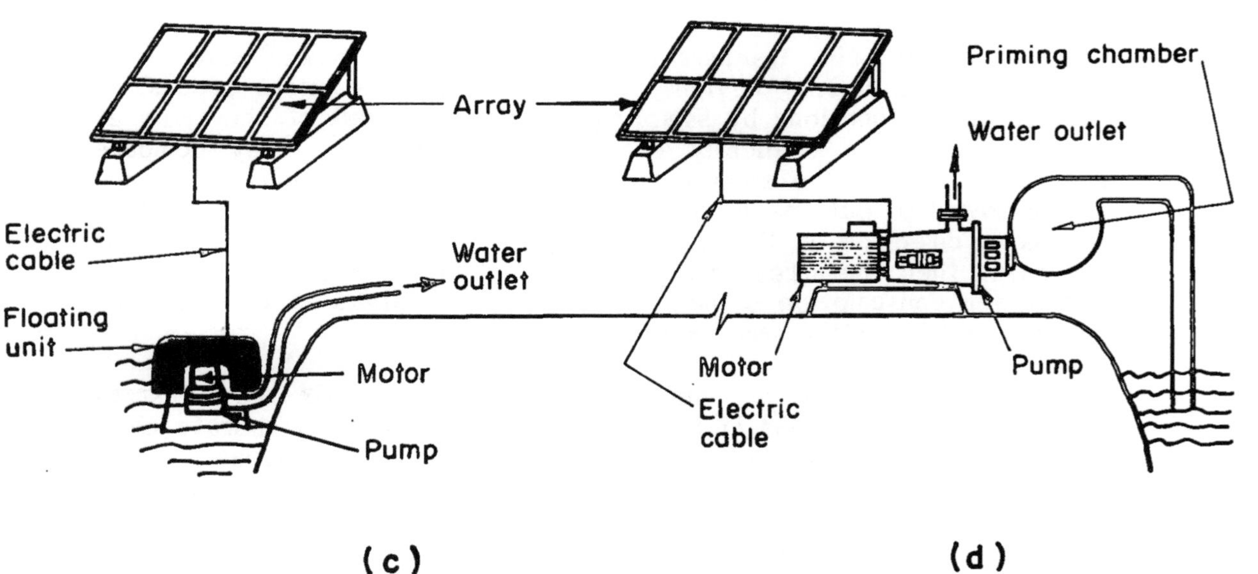


Figure 11. Examples of solar pump configuration: (a) submerged motor/pump set; (b) submerged pump with surface motor; (c) floating motor/pump set; (d) surface motor with surface mounted pump.

Batteries also provide impedance matching and allow the motor to start at low or zero irradiance levels. They provide energy storage and allow designers to accurately optimise the sub-system because they operate at a fixed voltage. However, they have several disadvantages: they involve a power loss, increase the risk of reliability problems, have a shorter operational life than the rest of the solar pump, and require regular maintenance. At present, most solar pumping systems do not include batteries although where water storage is required they may justify further consideration.

On-Off switches and devices to protect components against power surges are recommended for use with solar pumps. Safety considerations must not be neglected - for example, high voltages should be avoided.

### 2.3.5  Subsystem Configuration

There are several combinations of motor and pump that are suitable for use with solar power. Self priming is an essential requirement because of the frequent number of unattended starts which occur throughout the day as a result of cloud variations.

Figure 11 illustrates the four principal configurations that are used at present. These are:

> (a) submerged motor/pumps with centrifugal pumps, often consisting of several impellers and therefore termed "multi-stage"

> (b) a submerged pump arrangement with the pump below the water, driven by the shaft from a motor mounted above the water. Figure 11 shows a centrifugal pump, although this could equally be a positive displacement pump, usually in the form of a reciprocating double acting piston pump

> (c) floating motor/pump units with centrifugal pump

> (d) surface mounted pumps with a self priming tank. Positive displacement pumps have better self-priming characteristics.

Each configuration is suited to a particular range of head and flow.

Figure 12 shows which pump types are suitable for the different ranges of head and flow when using solar power.

### 2.3.6  Sub-system Efficiency

Table 1 shows typical sub-system types for alternative pumping applications. For system sizing purposes, and hence economic evaluation, it is important to know the daily energy efficiency and the peak power efficiency of the systems involved. Typical values obtained from tests undertaken for the UNDP/World Bank Solar Water Pumping Project are given in Table 1. The energy efficiencies are based on the total output from the array (including that below the pumping threshold irradiance levels) and hence are dependent on the solar irradiance profile for the day. For example if the peak solar irradiance is 500

| Lift | Sub-system type | Typical * subsystem daily energy efficiency | | Typical * subsystem peak power efficiency | |
|---|---|---|---|---|---|
| | | Average | Good | Average | Good |
| 2 metre | Surface Suction or floating units with submerged suction utilising brush or brush-less permanent magnet d.c. motors and cent-rifugal pumps | 25% | 30% | 30% | 40% |
| 7 metre | - Floating d.c units with sub-merged pump<br>- submerged pump with surface mounted motor, brush or brush-less permanent magnet d.c. motors single or multi- stage centrifugal pumps | 28% | 40% | 40% | 60% |
| 20 metre | - a.c. or d.c. submerged multi-stage centrifugal pump set, or<br>- submerged positive displacement pump with d.c. surface motor | 32% | 42% | 35% | 45% |

Table 1.  State of the Art for Motor/Pump Subsystems

Note
* calculated for a 20° array inclination, with a daily solar
  irradiation of 5 kWh/m$^2$, on a horizontal surface  with a peak
  solar irradiance of 708 W/m$^2$ and a diffuse fraction of 34%.
  For lower irradiation values the subsystem efficiencies are
  likely to be less.

W/m$^2$ and the cut-in threshold level is 400 W/m$^2$ then the daily energy efficiency will be significantly lower than the values given in Table 1.

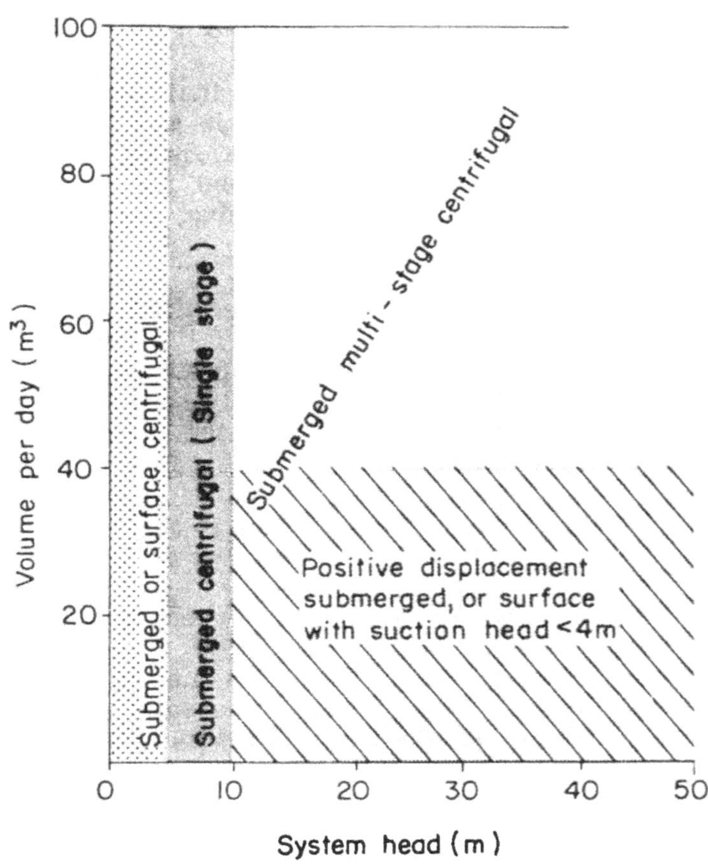


Figure 12.   Pump types suitable for a range of pumping applications.

## 2.4   Storage and Distribution

The main factors to consider in the delivery system are:

o   the volumetric efficiency of storage and distribution, which is defined as the proportion of pumped water which actually reaches its point of use – higher efficiency delivery systems can use smaller sized pumps

o   the pressure loss in the delivery system and the static head of the storage tank – lower head systems can use smaller sized pumps.

Often systems that are efficient in their use of water (such as pressurised-drip systems for irrigation) require high driving heads, and so to optimise the overall system performance, any increase in delivery efficiency should be weighted against the increase in driving head.

### 2.4.1 Storage of Water

One of the major problems with solar energy is that power is not available on demand. For irrigation applications it may be critical that water is available to prevent a crop from dying, and it is usually equally important to have water on demand for rural water supplies. Therefore, when using solar powered pumping, the problem of water storage must be considered.

For irrigation two types of water storage can be identified:

(a) long term storage in which water is stored from month to month to even out the demand pattern. This type of storage will permit irrigation on demand, and minimise the effect of variations in monthly water demand;

(b) short term storage which allows a farmer to store water from one day to the next. This serves the dual purpose of;

o giving the farmer improved water management control, and

o smoothing out the day to day variations, i.e. on a day with a high level of solar energy, when the solar pump could provide so much water that it would saturate the soil, the excess water would be stored so that it could be used for a day with a low level of solar energy.

Long term storage for irrigation systems is not usually feasible for practical and economic reasons, and to a certain extent, the soil itself can sometimes act as a short term store to even out the effect of good and bad days in a month. However, for most applications, the use of short-term water storage systems should be considered.

For Rural water supplies it is essential to include a storage tank when using solar pumps. Preferably this should meet several days water demand. Where there is a piped distribution system, the storage tank will have to be raised above ground level to provide sufficient pressure for the water to flow in the pipes.

Since increases in total water lift have a proportional effect on the cost of a solar pump, it is important that any storage tank should have a low aspect ratio (i.e ratio of height to diameter). It should also be remembered that water can be lost by evaporation and seepage if the tanks are not covered and lined, thereby decreasing the efficiency of the storage system. Tanks should always be covered to minimise the entry of dirt, insects and animals.

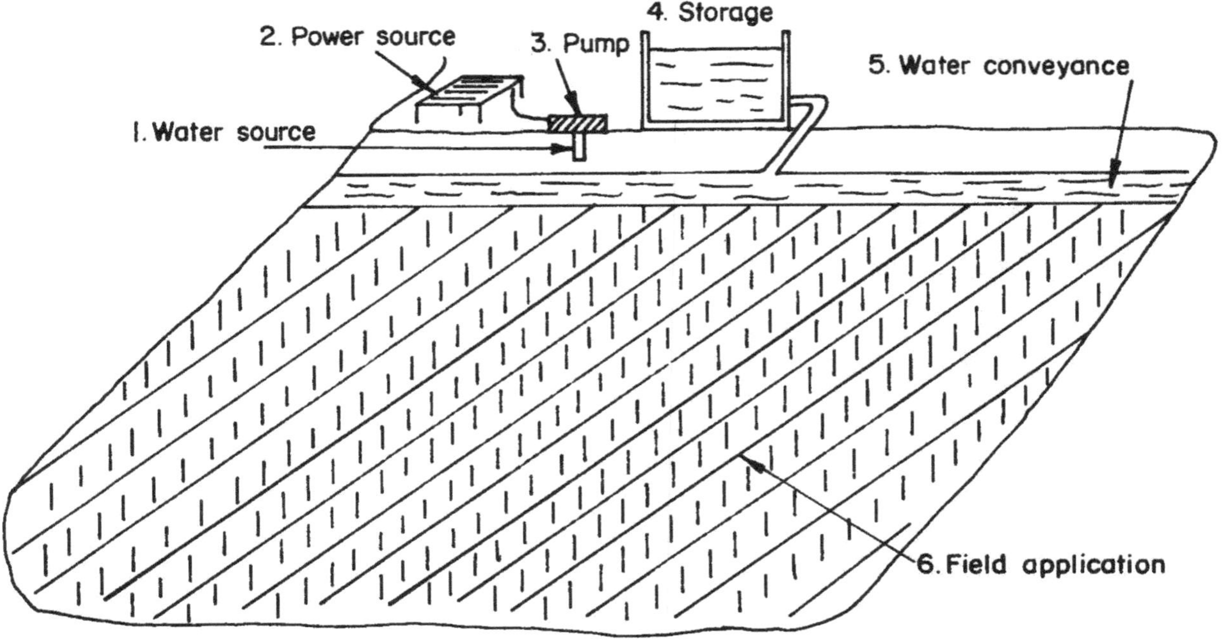


Figure 13.    Schematic layout of a small scale irrigation system, showing the six major components.

## 2.4.2  Distribution

### Irrigation

The distribution system for a small scale irrigation scheme consists of two parts: a water conveyance network to transfer water from the pump (or storage tank) to the field, and a field application method to apply water to the crops (Figure 13). The suitability for use with a solar pump of each of the four main methods of distribution is considered below:

Channel Irrigation: With channel irrigation the conveyance network will normally be in open channels. Losses are due to evaporation (usually small, around 1%), seepage, and evapotranspiration from weeds in the channel (can be high in earth channels, 30% to 50%). The additional head due to a conveyance channel depends on the slope and channel length. The field application method will normally be furrows. Losses of irrigation water from the furrows occur due to surface run off and deep percolation below the root zone. An overall distribution efficiency of 50%-60% is typical.

Trickle Irrigation: For this method water is conveyed in a main pipe and applied to the field continuously with small perforated trickle pipes. High irrigation efficiencies are possible, typically 85%. The head loss in the pipes is dependent on the water supply flow rate and the diameter of the pipes. Unless the area to be irrigated is level, the head loss has to be comparatively high to ensure an even water distribution, making the method less attractive for use with solar pumps.

Flood Irrigation: In this method the field is divided into small units and earth banks are constructed up to 30 to 50 cm high around each unit. The basins are filled to within 10 cm of the top of the banks. The size of the basins depends on the rate of water supply available. The main disadvantage here is the peaky demand and the high flow rate required. Since the size of the pump is determined by the peak capacity at flooding time, it is unlikely that solar pumps are suitable for flood irrigation, unless used in conjunction with another pumping device.

Sprinkler Irrigation: The efficiency of sprinkler irrigation is typically about 70%. The area watered by a given sprinkler depends on the operating pressure. For a diameter of coverage between 6 m and 35m the sprinkler operating pressure is usually between 1 and 2 bar, representing an additional head of 10 - 20 metres. This is not an energy efficient way of irrigation and consequently is not normally to be recommended for solar pumps.

Table 2 summarises the suitability of each method for use with solar pumps.

| Distribution Method | Typical Application Efficiency | Typical head | Suitability for use with solar pumps? |
|---|---|---|---|
| Open channels and furrows | 50 - 60% | 0.5 - 1m | Yes |
| Sprinkler | 70% | 10 - 20m | No |
| Trickle | 85% | 1 - 2m | Yes |
| Flood | 40 - 50% | 0.5m | No |

Table 2.   Suitability of major irrigation distribution methods for use with solar pumps.

<u>Rural Water Supply</u>

For many locations, distribution systems will not be considered because of the additional cost involved due to (a) the head loss in the distribution pipes and (b) the cost of installing the pipes.

A prime factor to consider when deciding on a distribution system for livestock or for a village water supply is the number of livestock or people to be supplied by one pump. There are few economies of scale with solar pumping, and it can therefore be expected that several pumps would have a similar cost to that of one pump with piped distribution. However, the overall system cost needs to be minimised for each application and due allowance made for the cost of drilling wells or boreholes.

A village water supply needs to be designed to suit the residents of the village. There are several factors affecting rural water use habits, all of which should be taken into account when designing a water supply system. For example the time required for collecting water should be considered when choosing both the number of water points and the distance between them. Figure 14 shows a schematic arrangement of a village water supply with distribution.

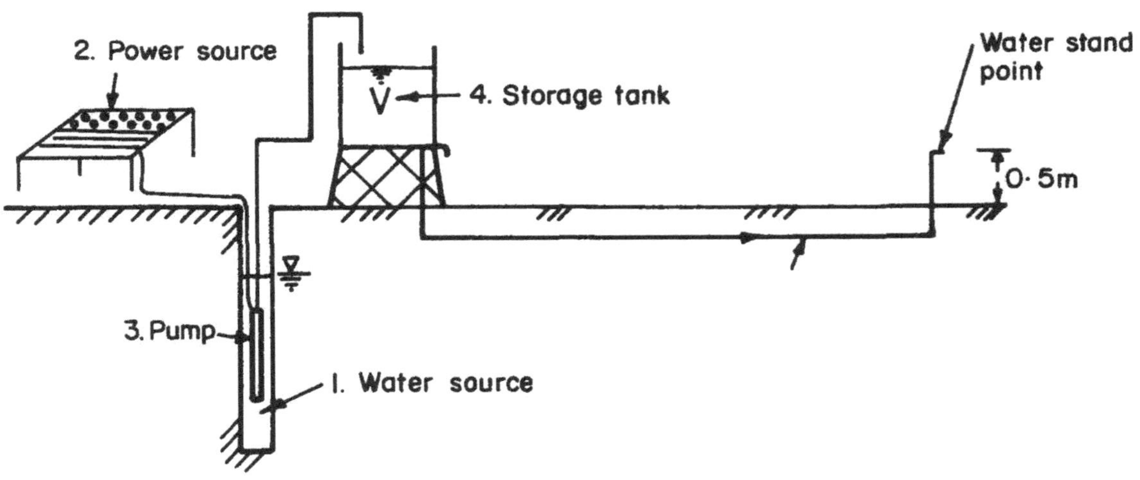


Figure 14.   Schematic layout of a village water supply system showing the five major components.

## 2.5  A Typical Day's Operation of a Solar Pump

To illustrate the concepts involved, Figures 15 and 16 show a typical clear sky daily irradiance pattern and the resulting power, energy and water flows, together with component efficiencies for a solar pump. The system uses a 250 Wp array, providing approximately 40 m$^3$ of water per day through a 5 metre lift. Two sub-systems are considered:

System A:    (Figure 15) d.c permanent magnet motor with centrifugal pump and no power conditioning. The subsystem has a peak power efficiency of 55%.

System B:    (Figure 16) a.c 3 phase motor with centrifugal pump and inverter incorporating a power conditioner. The subsystem has a peak power efficiency of 64%.

System A starts to pump at 8.00 a.m. when the solar irradiance exceeds the cut-in threshold of 300 W/m$^2$. The pumped output increases until noon. The PV array efficiency is relatively constant throughout the morning because temperature increases have the effect of reducing the PV array efficiency. The match between the array and subsystem improves as the sub-system approaches its design operating point. The subsystem efficiency is held constant from 10.00. a.m to 2.00. p.m because it is well matched to the array. In the afternoon further temperature increases and matching losses reduce the array efficiency, but as the array cools down in the late afternoon its efficiency increases. The system stops pumping at about 4 p.m. with a cut-out threshold of 300 W/m$^2$. The subsystem energy efficiency is 50% and the pump provides 40 m$^3$ during the day.

System B has a lower cut-in threshold (250 W/m$^2$) and starts pumping at 7.30 a.m. The PV efficiency gradually decreases until 11.00 a.m because of temperature increases. The PV efficiency is higher than for system A because of the impedance matching electronics in the inverter. The subsystem efficiency follows a similar pattern to System A. Although the daily subsystem energy efficiency is lower than for System A (because of the losses on the inverter), there is an overall improvement in water output.

**SYSTEM PROPERTIES**

Array rating = 250 Wp

Solar irradiation = 25 MJ/$m^2$

Array electrical outputs = 3·9 MJ

Sub-system energy efficiency = 50%

Total head = 5 m

Water output = 40 $m^3$

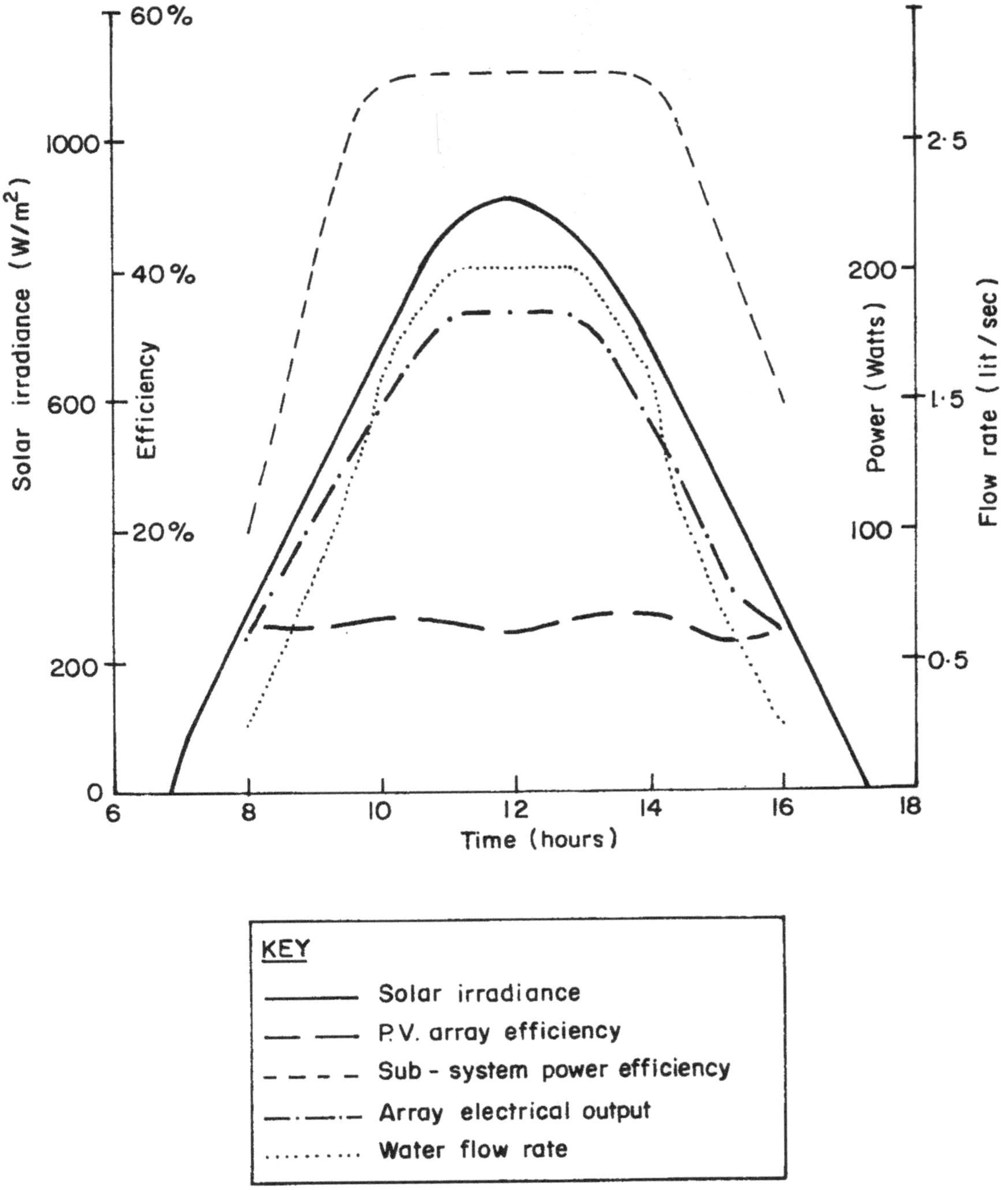

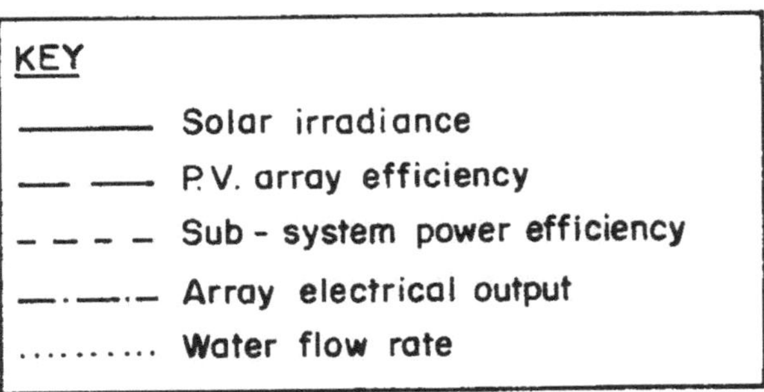

Figure 15. A typical day's operation of a solar pump without power conditioning.

**SYSTEM PROPERTIES**

Array rating = 250 Wp

Solar irradiation = 25 MJ/$m^2$

Array electrical output = 4·4 MJ

Sub - system energy efficiency 48 %

Total head = 5m

Water output = 42 $m^3$

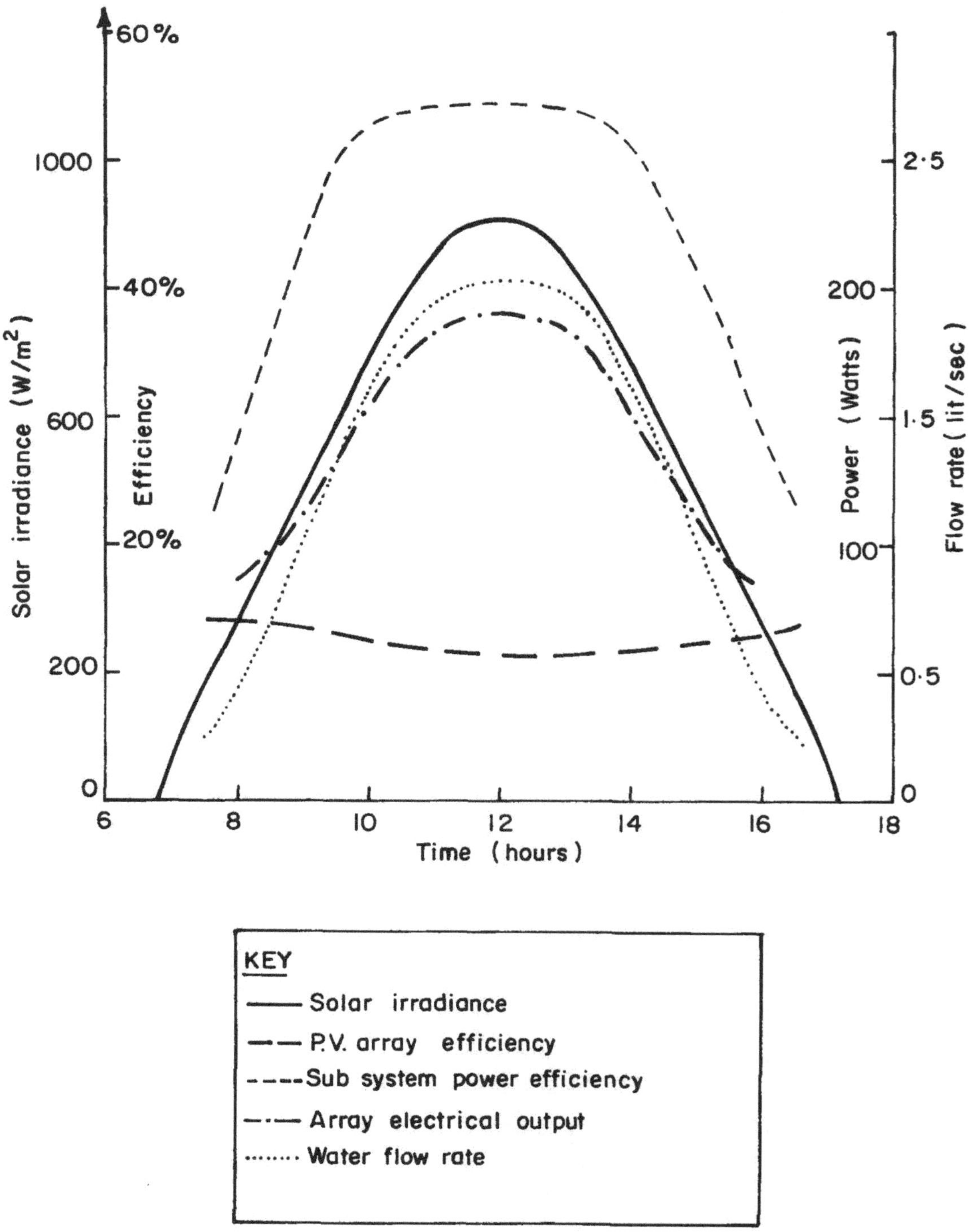

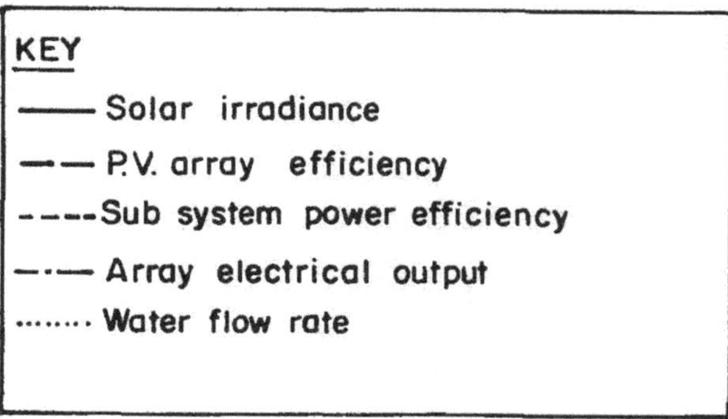

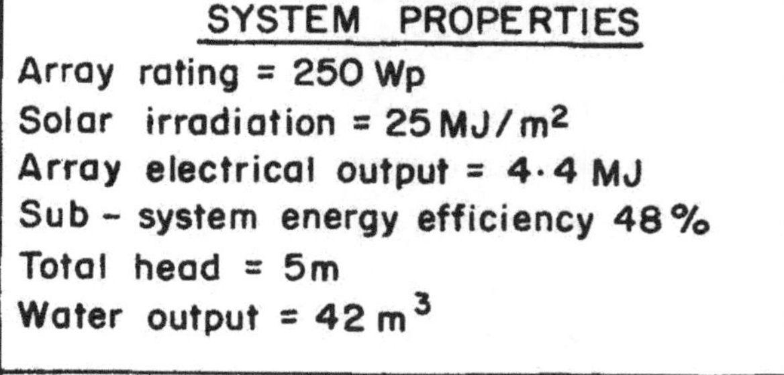

Figure 16. A typical day's operation of a solar pump with power conditioning.

# 3.  SITE EVALUATION

The end result of a site evaluation should be an answer to the question "What type and size of solar pump is required?". Once the solar pump has been sized, the system cost can be calculated and an economic comparison made with alternative pumping methods, as shown in Chapter 4.

There are a number of computer models available for sizing photo-voltaic and other pumping systems, but these vary widely in complexity and it is not the purpose of this publication to discuss computer models. The procedures given here have been developed for use by the non-specialist, and although they involve a number of approximations, they have the advantage that results can be obtained using only straightforward calculations and nomograms.

In this chapter it will be shown that a site evaluation can be approached in the following stages:

o    assessing water requirements
o    calculation of hydraulic energy required
o    determination of available solar energy
o    sizing the solar pump
o    selection of a suitable system configuration
o    specifying solar pump performance

For each stage a standard format sheet is given that can be used to carry out the calculations for a site evaluation. To illustrate the procedure the formats are completed for the example system detailed in Table 3. It is assumed that solar radiation data are not available for this site. Further examples of rural water supply systems are given in Appendix 4 and blank format sheets in Appendix 5.

| | |
|---|---|
| Location: | Bura, Kenya, Latitude $1^{\circ}$S |
| Application: | Irrigation |
| Area: | 1 hectare |
| Cropping Pattern: | Feb. to Mar. Crop 1:  Cotton 1 ha |
| | Sept. to Jan. Crop 2: Groundnuts (0.16 ha) |
| | Maize (0.32 ha) |
| | Oct. to Jan. Crop 3:  Cowpeas (0.32 ha) |
| Water Source: | Open Well |
| Static Lift: | 2m |
| Water conveyance network: | 30  metre PVC pipe, efficiency 100% |
| Field Application Method: | Furrow, efficiency 60% |

Table 3    Specification of Site for the Example Solar Pumping System

## 3.1. Assessing Water Requirements

### 3.1.1. Irrigation

The quantity of water needed to irrigate a given land area depends on
a number of factors, the most important being:

- o nature of crop
- o crop growth cycle
- o climatic conditions
- o type and condition of soil
- o land topography
- o field application efficiency
- o conveyance efficiency
- o water quality

Many of these vary with the seasons, and the quantity of water required is far from constant. The design of a small irrigation pump
installation will need to take all these factors into account.

The crop takes its water requirements from moisture held in the soil. Useful water for the crop varies between two levels the: "permanent
wilting point" and "field capacity" (see Figure 17). Water held by
the soil between these two levels acts as a store. When this store
approaches its lowest level, the crop will die unless additional water
is supplied.

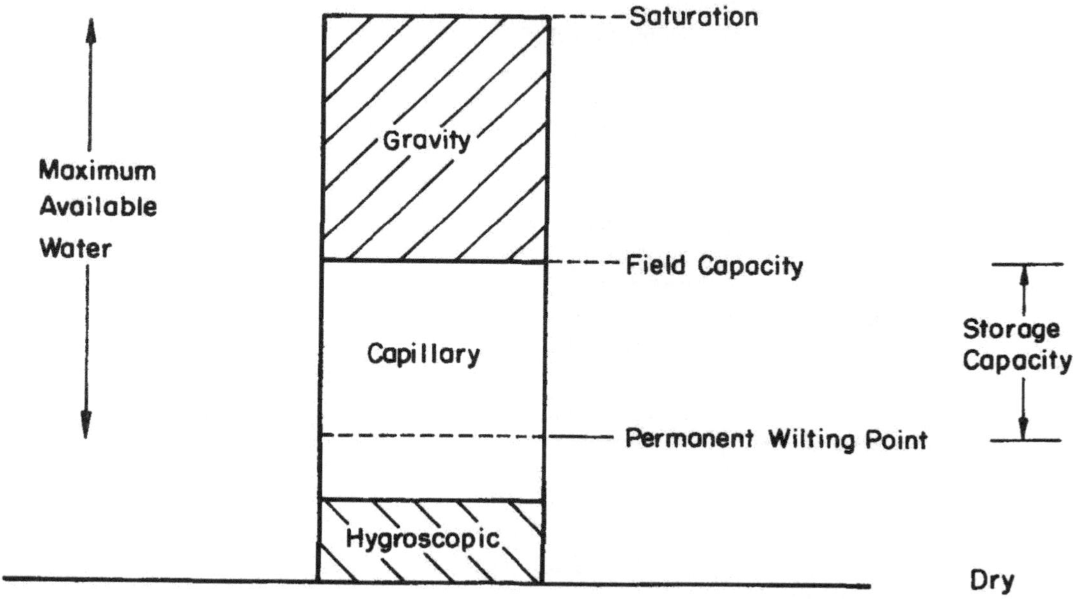


Figure 17.    Soil moisture quantities.

The rate of crop growth depends on the moisture content of the soil. There is an optimum growth rate condition in which the soil water content lies at a point somewhere between the field capacity and the permanent wilting point (Figure 18). However, this point varies for different crops and for different stages of growth and so it is not easy to adjust the irrigation intervals so that there is optimum crop growth.

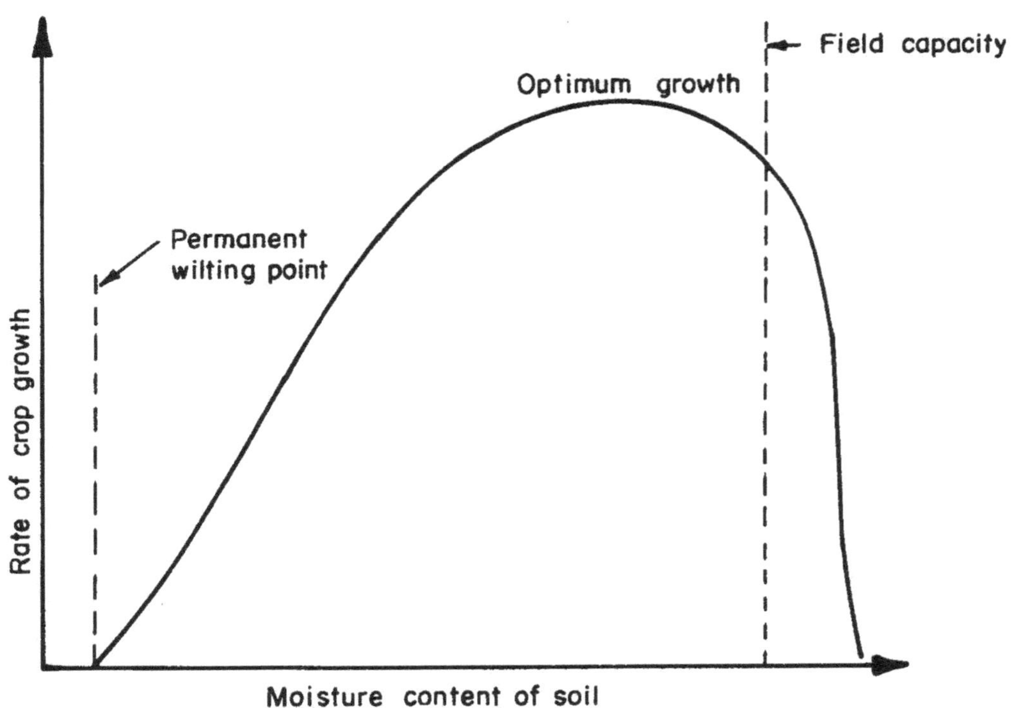


Figure 18.    Rate of crop growth as a function of soil moisture content.

An estimation of the quantity of water that is required for irrigation can usually be obtained from local experts and agronomists. It involves several calculation stages [†]:

> (i) Prediction methods are used to estimate crop water requirements, because of the difficulty of obtaining accurate field measurements.

> (ii) The effective rainfall and groundwater contributions to the crop are subtracted from the crop water requirements to give the net irrigation requirements.

------------------------------------------------------------------------

[†] For full details of the calculation procedures for irrigation water requirements refer to "Crop Water Requirements" by J. Doorenbos and W.O. Pruitt, FAO, Rome, (1977).

(iii) Field application and water conveyance efficiency are taken into account to give the gross pumped water requirements.

To illustrate the variation in irrigation water requirements between different crops and different locations, Table 4 gives water requirements for alternative crops in Bangladesh and Thailand. Irrigation water requirements for the example system are shown in Table 5.

| Month | Bangladesh<br>April to July: Rice<br>Oct. to April: Vegetable<br>$m^3$ per day per hectare | Thailand<br>Jan to Dec: Sugar Can<br><br>$m^3$ per day per hectare |
|---|---|---|
| January | 7.1 | 1.3 |
| February | 17.5 | 27 |
| March | 28.4 | 32 |
| April | 85.0 | 42 |
| May | - | 42 |
| June | - | 31 |
| July | - | 28 |
| August | - | 22 |
| September | - | 12 |
| October | - | - |
| November | 15 | - |
| December | 16.5 | 21 |

Table 4.    Typical Irrigation Water Requirements for Bangladesh and Thailand

| Location **Bura**   Latitude $1^{\circ}$S   Area **1** ha | | |
|---|---|---|
| Distribution method **Pipe & furrow**   at **60 %** efficiency | | |
| Month | Net (crop) water<br>requirements<br>$m^3$/day | Gross (pumped) water<br>requirements<br>$m^3$/day |
| January | 2.5 | 4.2 |
| February | 50 | 83 |
| March | 53 | 88 |
| April | 44 | 73 |
| May | 60 | 100 |
| June | 56 | 93 |
| July | 37 | 63 |
| August | 28 | 47 |
| September | 25 | 42 |
| October | 22 | 37 |
| November | 24 | 40 |
| December | 20 | 35 |

Table 5 Irrigation Water Requirements for the example system

### 3.1.2.  Rural Water Supply

The estimation of water demand for villages and livestock is considerably easier than the procedure for irrigation, because the volume
required is simply the product of population and per capita consumption. The discussion below indicates typical values that can be used.

Domestic water requirements per capita vary markedly in response
to the actual quantity of water available. For example the western
world uses several hundred litres per capita per day. In developing
countries, if there is a house supply, the consumption may be five or
more times greater than if water has to be carried from a public water
point.

A WHO survey in 1970 showed that the average water consumption in
developing countries ranges from 35 to 90 litres per capita per day. The
long term aim of water development is to provide all people with ready
access to safe water in the quantity they want. However, in the short
term a reasonable goal to aim for would be a water consumption of
about 40 litres per capita per day. Thus for typical village populat-
ions of 500, water supplies will have to be sized to provide about 20
$m^3$ per day. As discussed in Chapter 2 there are limits to the number
of people who should be served by one centrally located water point.
In order not to cause unreasonable water collecting times and carrying
distances, a single pump or water point should usually supply no more
than about 500 people.

Table 6 shows typical daily water requirements for a range of
livestock.

| Animal | Water Requirement litres per day |
|---|---|
| Horse | 50 |
| Dairy Cattle | 40 |
| Steers | 20 |
| Pig | 20 |
| Sheep | 5 |
| Goat | 5 |
| Poultry | 0.1 |

Table 6.    Typical Daily Water Requirements for Livestock


To prevent overgrazing in the locality of water points, the water
point should be kept reasonably small. For example, for cattle each
water point might serve 500 animals giving a total requirement of 20
$m^3$ per day.

### 3.1.3. Hydraulic Energy Requirements

Once the gross water requirements are known, the hydraulic requirements can be determined, as outlined in Chapter 1, using the equation:

$$\text{Hydraulic energy} = \frac{9.81 \times \text{volume (m}^3 \text{ per day)} \times \text{total head (m)}}{1000}$$

The total head comprises the sum of the static lift (including that due to the delivery and storage system) and the head loss in any pipes, which depend on the pipe diameter and the flowrate as shown in Figure 21. The flowrate will vary throughout the day and depend on the size of the pump (or power source). Since the power source size depends on the total head, the procedure to calculate pipe head loss should be iterative. However, for most systems the variation in head loss throughout the day will be small and the pipe diameter may be chosen so that the head loss is a small proportion of the total head. For example the pump can be sized by assuming that the pipe head loss is 10% of the total static head. Then, once the pump has been sized, the peak flow rates can be used to determine how large the pipes must be to give this assumed head loss. (Note if, after carrying out this procedure, the pipe sizes are found to be unreasonably large, the procedure can be repeated with a higher percentage head loss. If pipes are not included in the system, the static head alone may be used to determine the hydraulic energy requirements.)

Table 7 gives a format sheet for estimating hydraulic energy requirements which has been completed for the example system, assuming a 10% head loss in the pipes. In this case the static head is assumed to be constant throughout the year. However, where there are variations due to drawdown, monthly mean values of the static head should be used.


## 3.2 Determination of Available Solar Energy

Month by month solar radiation data are required, in order to assess adequately the suitability of a location for solar pumps. It is not sufficient to size a solar pump on the basis of annual solar energy availability because a pump sized in this way may not provide sufficient water in months of low solar insolation. For convenience, the monthly solar radiation is usually expressed in terms of the daily average irradiation for the month, i.e. $MJ/m^2$ per day. The radiation data that are available are usually in the form of global irradiation and these are the starting point for assessing the site. If possible, data should be obtained from the nearest available meteorological station, and due allowance made for any known localised micro-climate.

It is difficult to generalise on the accuracy of solar radiation data since numerous factors such as type and position of the measuring device used, need to be considered. Any errors in the radiation data will lead to proportional errors in the size of the photovoltaic array calculated in section 3.3, and hence in the unit water cost calculated using the procedure in Chapter 4. For most purposes a good 'rule of the thumb' is that the radiation data is accurate to within

Location ... BURA ........       Latitude .....1°S........

Delivery pipe head loss .10. %;   Delivery pipe length 30. m

| Month | Pumped volume requirement (m³/day) | Static head (m) | Dynamic head loss (m) | Total head (m) | Hydraulic energy (MJ/day) |
|---|---|---|---|---|---|
| Jan | 42 | 2 | 0·2 | 2·2 | 0·09 |
| Feb | 83 | 2 | 0·2 | 2·2 | 1·79 |
| March | 88 | 2 | 0·2 | 2·2 | 1·90 |
| April | 73 | 2 | 0·2 | 2·2 | 1·57 |
| May | 100 | 2 | 0·2 | 2·2 | 2·13 |
| June | 93 | 2 | 0·2 | 2·2 | 2·00 |
| July | 63 | 2 | 0·2 | 2·2 | 1·38 |
| Aug | 33 | 2 | 0·2 | 2·2 | 0·71 |
| Sept | 32 | 2 | 0·2 | 2·2 | 0·69 |
| Oct | 37 | 2 | 0·2 | 2·2 | 0·80 |
| Nov | 40 | 2 | 0·2 | 2·2 | 0·88 |
| Dec | 35 | 2 | 0·2 | 2·2 | 0·75 |

Table 7. Format sheet for calculation of hydraulic energy requirements

+ 10%. The reader may choose to oversize the photovoltaic array by this amount in order to allow for the uncertainty of the radiation data.

Where no local solar radiation data are available, an estimate can be made from the maps given in Appendix 1. These maps show the fraction of the extra-terrestrial solar energy that is transmitted to ground levelfor each month (this fraction is known as the clearness index) and have been prepared by the World Meteorological Organisation*.

---

* Meteorological Aspects of Solar Radiation as an Energy Source. World Meteorological Organisation. Technical Note No. 172. Geneva (1981).

To estimate the solar irradiation for a particular location one simply multiplies the extra-terrestrial solar energy for the location given in Table A1 by the clearness index for the location. Since the clearness index is only specified at intervals of 0.1, the accuracy of the resulting solar irradiation will be no better than $\pm$ 10%.

The solar radiation available on a tilted or tracking surface differs from that on a horizontal surface, and it is the solar radiation that the PV array receives that is important for the sizing procedure. Conversion factors must therefore be used to determine the irradiation on the array from the horizontal irradiation data. Unfortunately the formulae used to calculate these conversion factors are fairly complex and are tedious to use without a computer. The conversion calculations are also dependent on the fraction of diffuse irradiation. As a simplified procedure, Tables A2 to A10 have been prepared to estimate how the radiation on tilted surfaces is related to the horizontal irradiation. These Tables show the ratio of the solar irradiation on surfaces of different orientations to the solar irradiation on the horizontal plane as a function of latitude, month and clearness index†.

The procedure for calculating the solar energy availability is illustrated using the example system and the format sheet shown in Table 8. It is assumed that the PV array is at an inclination of $10°$ to the horizontal. (A horizontal array would receive more solar energy over the year but would not permit rain to run off). For example, in the month of May the extra-terrestrial irradiation at latitude $0°$ on a horizontal surface is 34 MJ/m$^2$/day from Table A1. From Figure A5 the clearness index at $1°$S, $40°$E is found to be 0.5 giving a global irradiation at the location of (34 x 0.5) = 17 MJ/m$^2$ per day. The tilt factor for the array at $10°$ from the horizontal is obtained from Table A2; for a clearness index of 0.5 and the month of May the tilt factor is 0.92, giving a solar irradiation on the PV array of 16 MJ/m$^2$ per day (4.4 kWh/m$^2$) per day.

## 3.3. Estimating Approximate System Sizing

Most solar pump suppliers will size the pump for a given location and water demand, but it is advisable to carry out an economic assessment of solar pumping viability, and to estimate the required system size before inviting proposals from suppliers.

To use the following procedure the month by month hydraulic energy requirement and solar energy availability must be known. There are two options for operating the solar pump: (i) as a stand alone system which must be sized to provide all the water requirements, or, (ii) as a fuel (or labour) saver in which the solar pump provides a basic water requirement, but where peak demands are met by an alternative method.

---

† The Tables have been calculated for an average day of the month using a correlation relating the diffuse irradiation to the clearness index.

<table>
<tr><td colspan="6" align="center">SOLAR ENERGY AVAILABILITY</td></tr>
<tr><td colspan="6">Location .BURA... Latitude ....1°S.... Array Tilt ..10°......<br>Longitude ...40°E...</td></tr>
<tr>
<td>Month</td>
<td>Extra-terrestrial Irradiation MJ/m²</td>
<td>Clearness Index</td>
<td>Global Horizontal Irradiation MJ/m²</td>
<td>Tilt Factor</td>
<td>Global Irradiation on Array MJ/m²</td>
</tr>
<tr><td>Jan</td><td>38</td><td>0.3</td><td>21.6</td><td>1.08</td><td>23</td></tr>
<tr><td>Feb</td><td>37</td><td>0.3</td><td>22.2</td><td>1.04</td><td>23</td></tr>
<tr><td>March</td><td>38</td><td>0.6</td><td>22.8</td><td>1.00</td><td>23</td></tr>
<tr><td>April</td><td>36</td><td>0.5</td><td>18.0</td><td>0.95</td><td>17</td></tr>
<tr><td>May</td><td>34</td><td>0.5</td><td>17.0</td><td>0.92</td><td>16</td></tr>
<tr><td>June</td><td>33</td><td>0.5</td><td>17.5</td><td>0.90</td><td>16</td></tr>
<tr><td>July</td><td>34</td><td>0.5</td><td>17.0</td><td>0.91</td><td>15</td></tr>
<tr><td>Aug</td><td>35</td><td>0.5</td><td>17.5</td><td>0.94</td><td>16</td></tr>
<tr><td>Sept</td><td>37</td><td>0.6</td><td>22.2</td><td>0.98</td><td>22</td></tr>
<tr><td>Oct</td><td>37</td><td>0.5</td><td>18.5</td><td>1.02</td><td>19</td></tr>
<tr><td>Nov</td><td>36</td><td>0.5</td><td>18.0</td><td>1.06</td><td>19</td></tr>
<tr><td>Dec</td><td>35</td><td>0.5</td><td>17.5</td><td>1.07</td><td>19</td></tr>
</table>

Table 8.   Format for calculation of solar energy availability

The sizing methodology for both cases is identical once the design month has been established. This is the month in which the water demand is highest in relation to the solar energy available, i.e. the month when the system will be most heavily loaded to meet the demands. For the stand alone system the design month is found by calculating the ratio of the hydraulic energy requirement to the solar energy available for each month. The month in which this ratio is a maximum is the design month. For a fuel saving system, the baseline requirement needs to be established. For example, this could be the average water requirement throughout the year. Then the ratio of the baseline hydraulic energy to the solar energy available can be calculated for each month. Again the month with the maximum ratio is the design month.

The data for the design month are used to calculate the required component sizes in the step by step procedure given below. Table 9 shows a format sheet for making the calculations. Each step is illustrated for the example system. In this case the design month is May with a hydraulic energy requirement of 2.16 MJ/day (0.6 kWh/day).

Step 1: **Size the PV array**

The electrical energy required from the PV array is equal to the required hydraulic energy divided by the average sub-system daily energy efficiency. The electrical output of the PV array depends on three factors (the latter two of which affect the array efficiency).

        o   the solar irradiation incident on the array

        o   the average cell temperature which in turn depends on ambient air temperature and solar irradiance levels.

        o   the electrical load because this determines the operating point on the PV array current/voltage (I/V) curve. For a solar pump without impedance matching electronics, the electrical output of the array is reduced below its maximum value except when operating at the knee of the I/V curve.

A detailed methodology for sizing a PV array is given in Appendix 3. The objective of the procedure is to determine the required array rating in peak watts (Wp). The principle of the method can be illustrated by first considering an array that is operating both at the reference cell temperature (of 25 $^{\circ}$C) and at the maximum power point on the current/voltage curve throughout the day. This means that when the solar irradiance is at a 1000 W/m$^2$ the PV array will produce its rated output.

The daily solar irradiation can be considered in terms of peak irradiance conditions at 1000 W/m$^2$ for an equivalent time period. For example, a daily irradiation of 18 MJ/m$^2$ (5 kWh/m$^2$) could be considered as equivalent to 1000 W/m$^2$ for a period of 5 hours. By assuming, as a first approximation, that the array will work at its rated output for this time period, then a first estimate of the array size can be made. For example, using this approximation, an electrical energy demand of 9 MJ per day (2.5 kWh) would require a PV array sized at 500 Wp, on a day with an irradiation of 18 MJ/m$^2$.

Under actual conditions the incident solar <u>energy</u> would be spread out over the daylight hours and the average <u>power</u> output from the PV array would be considerably less than the rated output. Also, in real conditions the array rating calculated above would be too small because of cell temperature effects and impedance matching losses. Therefore, it is necessary to increase the array rating by factors which account for the decrease in efficiency when not operating at reference conditions.

### SOLAR PUMP SYSTEM SIZING

Location ..BURA.. Latitude ...1°S... Longitude ..+0°E...

Design Month   MAY

Design Month hydraulic energy requirement ...2:18........ MJ

Design Month head ..............................2:2..... metres

Design Month global irradiation on array ......18....... MJ/m$^2$

Average sub system energy efficiency .................30..... %

Peak sub system power efficiency .................+0..... %

| Step | Calculations |
|---|---|
| 1. PV array size | Required electrical energy ....7.2..... MJ<br>PV array size .....................5+0.....Wp |
| 2.  Motor size | Rated motor input power ......5+0.... Watts |
| 3.  Pump size | Rated peak hydraulic power ...215... Watts<br>Rated flow rate ...............10... lit/sec<br>at .2:2.. metres head |
| 4.  Pipe Diameter | Diameter ..150... (mm) for head loss of<br>..0:2.... (m) at rated peak flow rate |

Table 9 Format sheet for calculation of system size

To guide the reader who does not wish to make the detailed calculations shown Appendix 3, the nomogram in Figure 19 has been prepared and can be used to determine the required PV array size to meet the hydraulic energy load for the design month. The starting point is axis OB where the hydraulic energy is given in MJ per day. Moving anti-clockwise and picking an appropriate sub-system daily energy efficiency from Table 1, the required electrical load in MJ per day is given on axis OC. The array rating in peak watts (Wp) is then selected from axis OA for the appropriate design month solar irradiation.

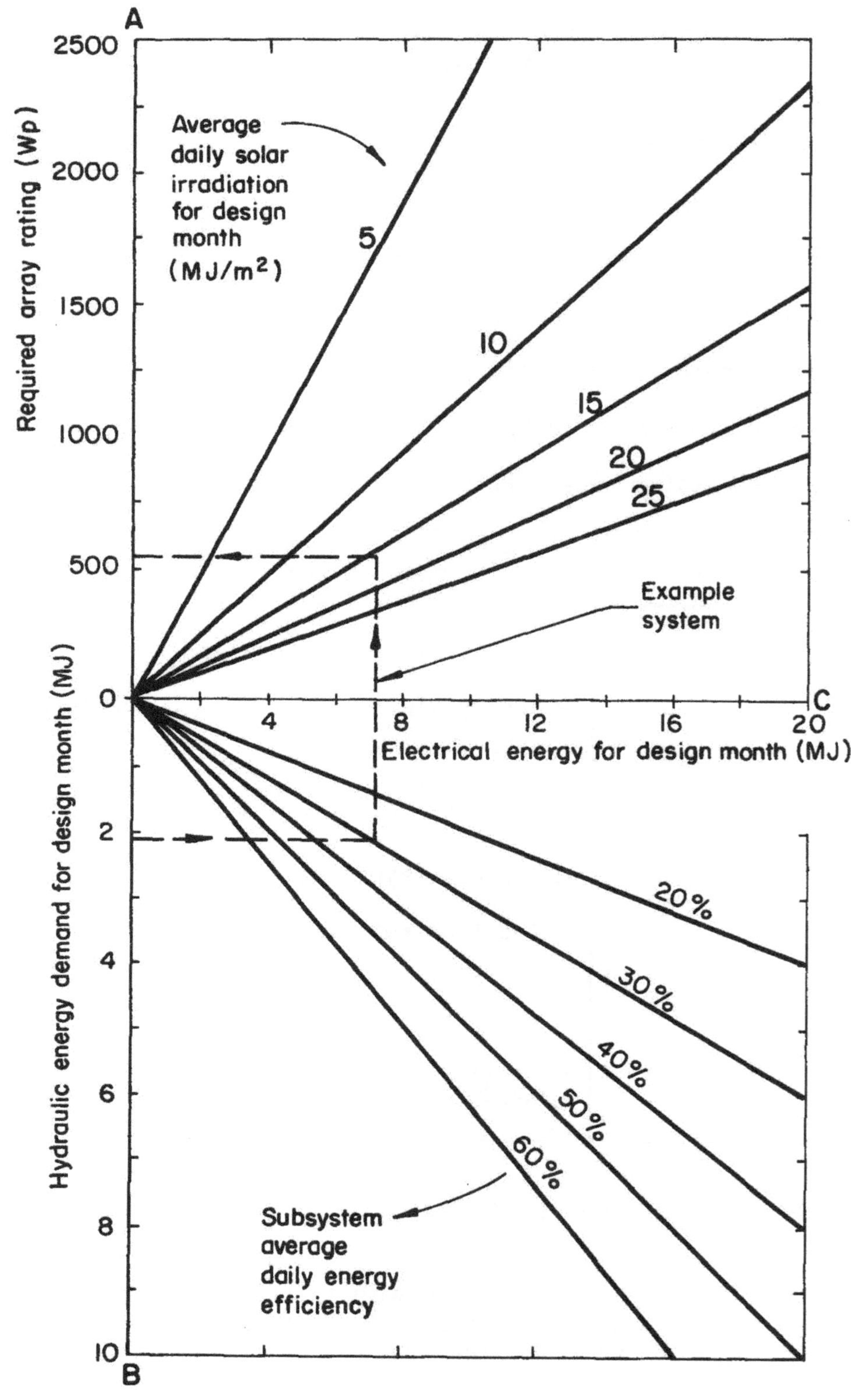


Figure 19.   Nomogram to determine the PV array rating for a given hydraulic duty

46

The array ratings shown have been calculated for a daily average cell operating temperature of 40°C and a system without power conditioning, which has a 10% loss in energy due to the mismatch between the PV array and the electric motor. If power conditioning is included, the required array rating would smaller than the values shown in Figure 19, but the reduction in size is unlikely to exceed 5 to 10%.

For the example system: (see dotted line on the nomogram). It can be seen from Table 1 that a system with a 2.2 metre lift will have a typical subsystem daily energy efficiency of 30%. The electrical energy requirement for an hydraulic energy demand of 2.16 MJ per day is found on axis OC to be 7.2 MJ. For a design month solar irradiation on the PV array of 16 MJ/m$^2$ per day, the array rating is found to be about 540 Wp.

## Step 2: **Size the motor**

The motor must be able to withstand the peak output of the array.
Since electric motors are generally rated in terms of their electrical input power, the maximum rating of the motor must be at least as great as the array rating. Thus the example system requires a motor rated at 540 Watts. The configuration of the PV array can usually be arranged to match the current and voltage limitations of the motor, provided that the maximum power ratings are adequate.

## Step 3: **Size the pump**

The peak hydraulic power output of the solar pump will be given by the product of peak array power output and peak subsystem power efficiency. Typical values of peak subsystem power efficiency are given in Table 1. The peak flow rate required from the pump can be obtained either by using the equation relating hydraulic power to flow rate and head (section 1.2) or by using the nomogram in Figure 20. The array rating is given on axis OB and the peak hydraulic power is obtained for an appropriate sub-system power efficiency. The peak flow rate can then be obtained from axis OA, for the required system head.

The dotted line in Figure 20 shows the path followed for the example system. With an array rating of 540 Wp and a peak subsystem power efficiency of 40% (obtained from Table 1), a peak hydraulic power of 215 Watts is found on axis OC, and a pump rated at a flow rate of 10 lit/sec for a head of 2.2 metres is found on axis OA.

## Step 4: **Size the Pipework (where included)**

The required pipe diameter to meet the head loss specified when calculating the hydraulic energy (Section 3.1.3), may be determined from Figure 21.

For the example system with a 10% head loss equal to 0.2m and a peak flow rate of 10 lit/sec, the required pipe diameter is 150 mm for a 30 m pipelength (i.e a head loss of 0.67 m per 100 metres of pipe). Note that if this pipe size were not available, the whole sizing procedure could be repeated using a greater percentage head loss. For example if 100 mm pipe was available the head loss would become 5 m per 100

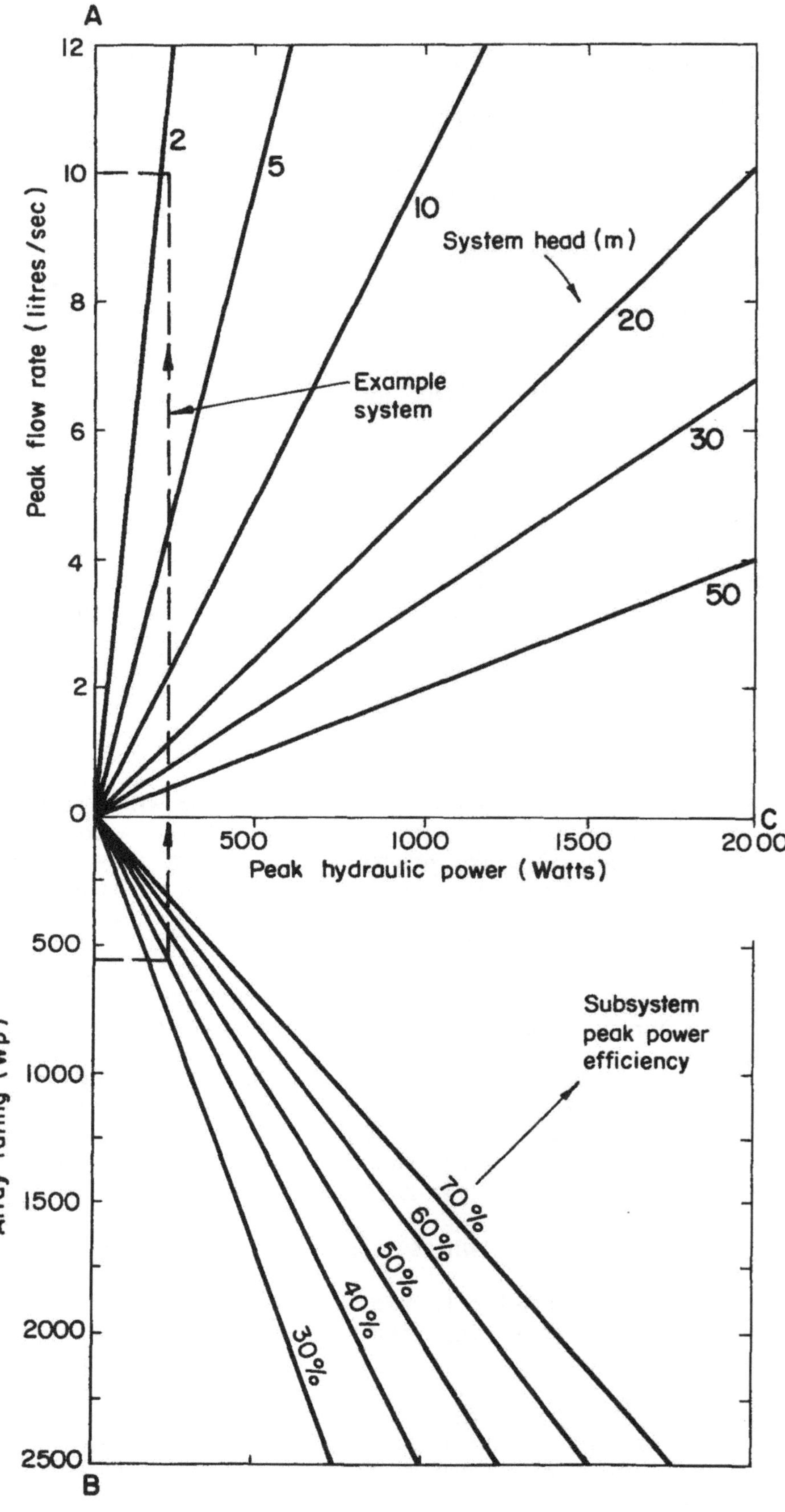


Figure 20.   Nomogram to determine pump rating for a given PV array rating

48

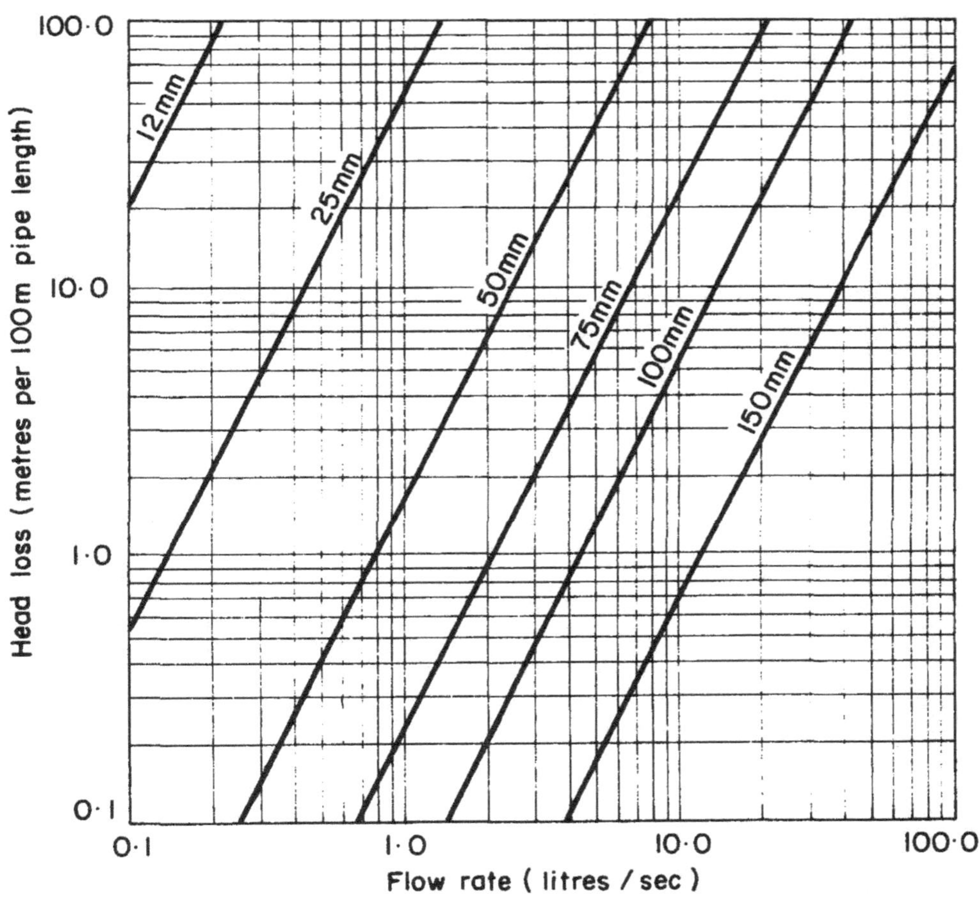



Figure 21. Head loss in smooth pipes of different internal diameter

metre of pipelength at a flow rate of 10 lit/sec. The total system head would then be 3.7 m and the calculations in sections 3.1.3 and 3.3 would need to be repeated.

## 3.4  Specification of System Performance and Configuration

The purchaser should now be in a position to make his/her own preliminary assessment of solar pumping viability in accordance with the Decision Chart discussed in Section 1.5, and to supply full details of his/her requirements. Before a purchase is completed it will be important to ensure that the purchased system is technically able to meet the demand and that it will meet the economic constraints.

A specification sheet, which may be included in a tender document (see Appendix 5), is shown in Table 10. When issuing the tender documents the purchaser need only complete parts 1-3 and the month by month pumped water requirements. However for the purchaser to make his or her own economic assessment before contacting a supplier, parts 4 to 6 should be completed as far as possible. These data are required for the economic assessment detailed in Chapter 4.

49

## SOLAR PUMP PERFORMANCE SPECIFICATION

Location ..BURA.. Latitude ...1°S... Longitude ..40°E...

1. Water source mean static lift .....................2........ m

2. Delivery system    Type .......PIPE & FURROW.........  
                              Length ..........................30...... m  
                              Pipe diameter ...............150.... mm  
                              Efficiency ...................60.... %

3. Storage system (when applicable)  
                              Volume ....................−........ $m^3$  
                              Height ....................−........ m

4. Design month details  
                              End use water requirement ..60.. $m^3$/day  
                              Pumped water requirement ..100.. $m^3$/day  
                              Hydraulic energy requirement 2.18 MJ/day  
                              Solar irradiation on  
                              PV array .............18.... MJ/$m^2$/day

5. Solar pump performance and water requirement

|  | Jan | Feb | Mar | Apr | May | Jun | Jul | Aug | Sep | Oct | Nov | Dec |
|---|---|---|---|---|---|---|---|---|---|---|---|---|
| Solar Irradiation ($MJ/m^2$/day) on PV array | 23 | 23 | 23 | 17 | 16 | 18 | 15 | 18 | 22 | 19 | 19 | 19 |
| Pumped water requirement ($m^3$ per day) | 42 | 33 | 38 | 73 | 100 | 93 | 63 | 33 | 32 | 37 | 40 | 35 |
| Pumped water ($m^3$ output/day) | 145 | 145 | 145 | 107 | 100 | 100 | 94 | 100 | 139 | 120 | 120 | 120 |

6. Solar pump specification  
                              PV array size ...............540... Wp  
                              PV array tilt ..................10°......  
                              Sub-system energy efficiency ...30... %  
                              Sub-system peak power efficiency .40. %  
                              Motor Rated Power .............540... W  
                              Pump rating .10. lit/sec at 2.2 m head

Table 10. Format sheet for Specification of Solar Pump Performance

The selection of a suitable subsystem configuration can be carried out with reference to Figure 12. For the example system with a system head of 2.2 m, either a submerged or surface centrifugal pump would be suitable.

It is not necessary to calculate the output in the remaining 11 months for stand alone systems because it is known that the solar pump can provide the annual demand (the pump has been sized to meet the 'worst' month water demand). However this calculation will give the over capacity of the pump, i.e. the amount of water that the pump can provide in excess of the requirements, and it may be possible to use the potential excess electrical output from the PV array to drive other loads. For solar pumps operating in an energy saving mode it is essential to calculate the output in the remaining months so that the unit water cost can be assessed.

The month by month water output can be obtained by using the nomogram in Figure 19 in the reverse direction. This will give the hydraulic output of the pump which can be converted to water output using the following equation:

$$\text{volume pumped} = \frac{1000 \times \text{hydraulic output (MJ/day)}}{9.81 \times \text{system head (m)}} \quad (\text{m}^3/\text{day})$$

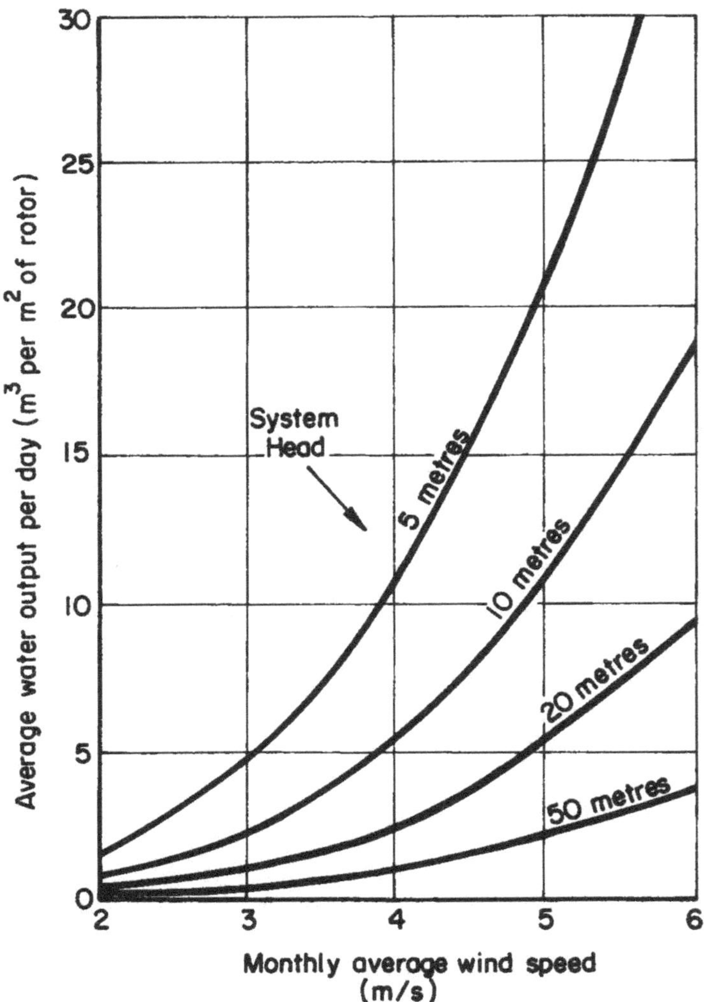


51

# 4.    ECONOMIC ASSESSMENT

## 4.1. Methodologies for Economic Evaluation

Economic considerations are important when comparing alternative pumping methods. Solar PV pumps are technically viable but where alternatives exist the evaluation of the alternatives must include both economic and technical considerations.

When economic viability is considered, a distinction should be made between economic and financial assessment. The economic approach seeks to make a true comparison of the value to society as a whole, and as such must use costs and benefits that are free from taxes, subsidies, interest payments etc. Conversely, a financial assessment is an evaluation from the purchaser's viewpoint; so taxes, subsidies and the effect of spreading the capital cost over several years by means of a loan, are all taken into account.

There are three common techniques that are used for making an economic appraisal:

> Payback Period: The length of time required for the initial investment to be repaid by the benefits gained is calculated.

> Rate of Return: The benefits gained are expressed as a rate of return on the initial investment.

> Life Cycle Costs: The sum of all the costs and benefits associated with the pumping system over its lifetime, or over a selected period of analysis, is expressed in present day money. This is termed the Present Worth (PW) of the system. For the system to be worthwhile, the benefits must be greater than the costs.

The most complete approach to economic appraisal is to use the life cycle costing because all future expenses are then taken into account.
In this method all the future costs and benefits are discounted to "present day" values. The underlying concept in this approach is that the investor would be indifferent as to whether he has $100 now or $110 in a years time if the $100 could be invested at an interest rate of 10%. Hence the Present Worth (PW) of an expenditure of $110 in one years time would be $100 when discounted at a rate of 10%.

The calculation of PW involves the use of a discount rate which reflects the opportunity cost of capital. Values of discount rate that are used for other projects in the country concerned can usually be taken as a guide, typical values are 10 - 12%. High discount rates mean that a low value is placed on future costs and benefits. For example at a discount rate of 50% an expenditure of $100 in one year's time has a PW of only $66.67.

For individuals the discount rate must reflect the purchasers view of the value of money available today, against money available in the

future. Again, high discount rates imply that the individual views
that money available at present is of more value.

<u>Calculation of the Present Worth.</u>

1. For a payment of Cr ($) to be made in the future the Present Worth
(PW) is found by multiplying the payment Cr, by a factor Pr:

$$PW = Cr. Pr$$

2. For a payment of Ca ($) occurring annually over a period of N
years, the Present Worth is found by multiplying the payment by a
factor Pa:

$$PW = Ca. Pa$$

The factors Pr and Pa depend on the rate of inflation (i), the discount rate (d) and the time (or period) of payment (N years). Formulae
and Tables relating Pa and Pr to these three parameters are given in
Appendix 2.

<u>Examples showing how to calculate the Present Worth</u>

> (a) To find the PW of a single cost of $10,000 occurring in 5
> years time at an inflation rate of 5% per annum and discounted
> at a rate of 10% per annum. The factor Pr from Appendix 2 for d
> = 0.1, i = 0.05 and N = 5 is found to be 0.79. Hence the PW =
> (10,000 x 0.79) = $7900.

> (b) To find the PW of an annual benefit of $2000 occurring for a
> period of 10 years which has an annual inflation rate of 5% per
> annum and is discounted at a rate of 15% per annum. The factor
> Pa from Appendix 2   for d = 15%, i = 5%, N = 10 is found to be
> 5.97. Hence the PW = (2000 x 5.97) =   $11,940.

<u>Note:</u>   It is usual to carry out economic evaluations in real terms;
the interest rates and discount rates used should be relative to
general inflation. Hence costs are only assumed to inflate or deflate
if their prices are changing relative to all other prices. However, as
long as both discount and inflation rates are expressed in the same
way (i.e. both excluding general inflation or both including general
inflation) the resulting Present Worth will be the same.


## 4.2.   Procedure for a Cost Appraisal of Water Pumping

Figure 22 outlines a methodology that can be used to compare the costs
of the major alternative pumping methods:

> o solar pumps
> o windpumps
> o engine driven pumps (diesel or kerosene)
> o animal powered pumps
> o handpumps

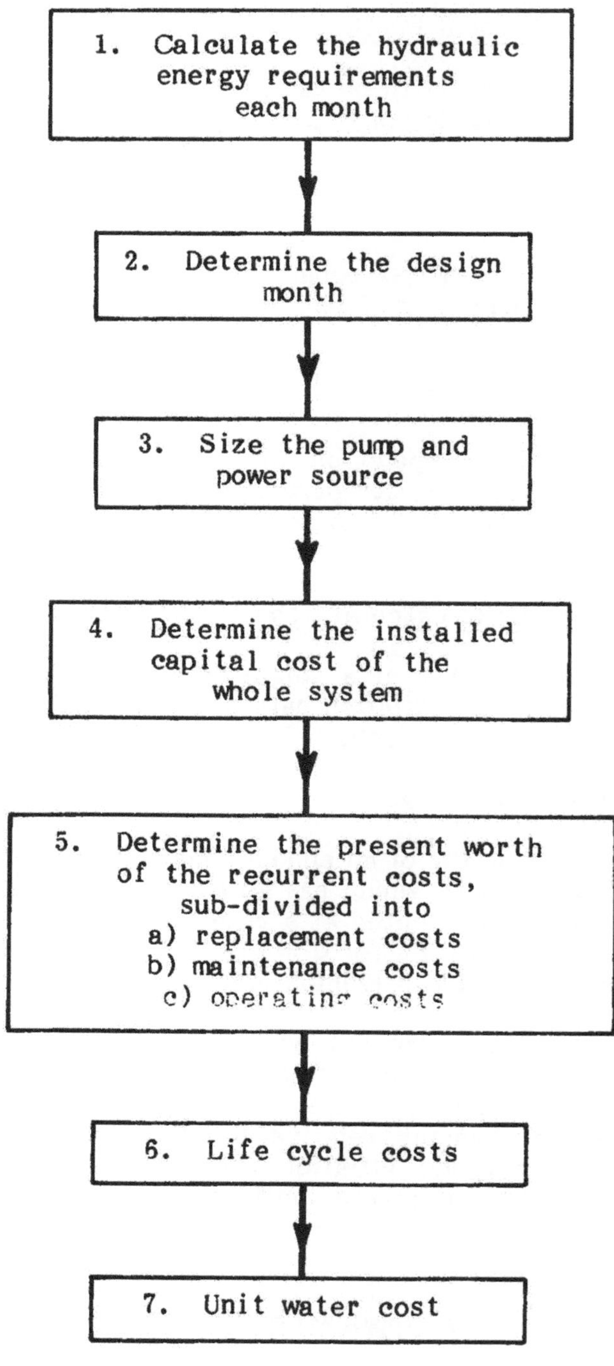


Figure 22.   Step by step procedure for a cost appraisal of a water pumping system.

This step by step procedure is based on a life cycle costing of the whole system. It takes into account each of the identifiable costs, but ignores the benefits gained by the users of the water. Consequently the results do not indicate whether a water pumping system is economically viable per se (for example whether additional crops grown using water supplied for irrigation are worth more than the cost of the water provided). However the methodology can be used to identify the pumping system which has the lowest life cycle cost and hence will provide the lowest unit water cost. Of course the least cost solution may not be the final choice since factors other than cost should also be taken into account. Reliability is of key importance and a user may be prepared to pay an increased cost for greater reliability. However a cost appraisal is a necessary step before making the final choice.

An integrated approach is used for the cost appraisal suggested here, considering the system as a whole from the water source to the point of use. In this way the influence of the storage and distribution systems on the overall unit water cost can be assessed. Figures 13 and 14 show schematic representations of typical irrigation and rural water supply systems.

For the purposes of the cost appraisal:

An <u>irrigation system</u> may be assumed to consist of six components; water source, power source, pump, storage tank (where necessary); water conveyance network; field application method.

A <u>rural water supply system</u> may be assumed to consist of a maximum of five components; water source, power source, pump, storage tank, and (where included) a piped distribution system.

The step by step cost procedure is described below. It is illustrated by calculating the unit water cost for the solar pumping system used as an example in Chapter 3. A format sheet for the cost appraisal is given in Table 11. Further examples comparing solar pumping with alternatives, are given in Appendix 4.

The first three steps relate to sizing the system and calculating the system energy requirements such that the costs can be found. The data required are the total system head, the month by month water demand and the appropriate meteorological data (solar irradiation or mean wind speed).

Step 1. **Calculate Hydraulic Energy Requirements Each Month**

The procedure is straightforward and was described in Section 3.1. Note that the total system head is the sum of: the distance between the level of the water source and ground level, the storage tank height and the head loss in the distribution system.

The hydraulic energy requirement is then determined from the product of the total head and the volumetric demand as outlined in Section 3.1.3.

<table>
<tr><td colspan="3" align="center">UNIT WATER COST FOR A SMALL SCALE PUMPING SYSTEM</td></tr>
<tr><td colspan="3">

1. System Description

Location..*BURA*.....     Latitude....*1°S*......

Design Month..*MAY*...     System Head..*2·2*.....m

Design Month water requirement*100*..m³/day  Power Source..*SOLAR*..

Annual water requirement.*12410*...m³   Size....*540 Wp*.....

</td></tr>
<tr><td colspan="3">

2. Cost Analysis

Period of analysis....*15*......years   Discount Rate..*10%*...

Pa......*7·61*............

</td></tr>
</table>

| | Annual Cost | Present Worth |
|---|---|---|
| Capital Cost | | **$7570** |
| Replacements | | **$769** |
| Maintenance | **$80** | **$609** |
| Operating | - | - |
| Total Annual Cost | **$80** | |
| Life Cycle Cost (LCC) | | **$8947** |
| Annual equivalent of LCC (ALCC) | | **$1176** |
| Unit Water Cost. | **9·5 cents / m³** | |

Table 11.   Format to calculate the unit water cost of a small scale pumping system.

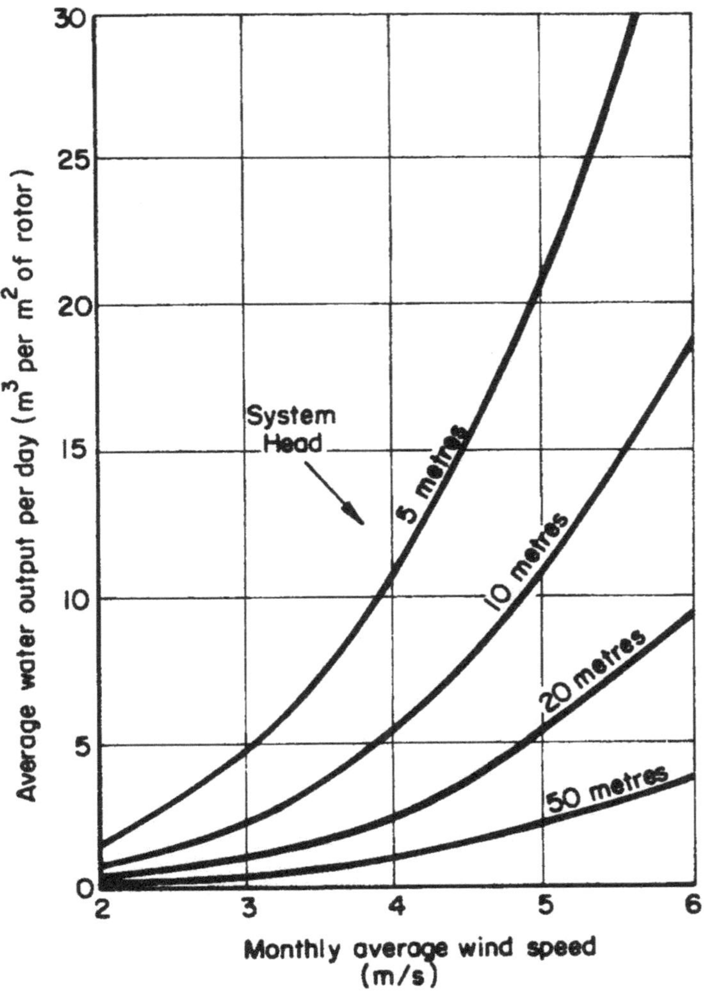


Figure 23. Average daily water output for windpumps expressed in
m³ per day per m² of rotor area. These curves can be used to obtain
the rotor area required for a given duty. They have been derived by
assuming the average (over 24 hours) hydraulic power developed by a
windpump is 0.1 x (windspeed)³ per square metre of rotor area.
Example 2, Appendix 4 indicates their use.


## Step 2. Determine the Design Month

A procedure for identifying the design month for solar pumps is outlined in Section 3.3. A similar procedure can be adopted for windpumps. Figure 23 shows the water pumped per square metre of swept
rotor area as a function of monthly average windspeed and total head.
These curves have been derived by assuming an average windpump performance; where available actual performance data should be used.

To determine the design month, the volume pumped per square metre of
swept rotor area must be determined for each month. The month with the
highest ratio of water requirement to pumped volume per m² of swept
rotor area is the designated design month.

For handpumps, animal and diesel pumps the design month is simply the
month with the highest water demand.

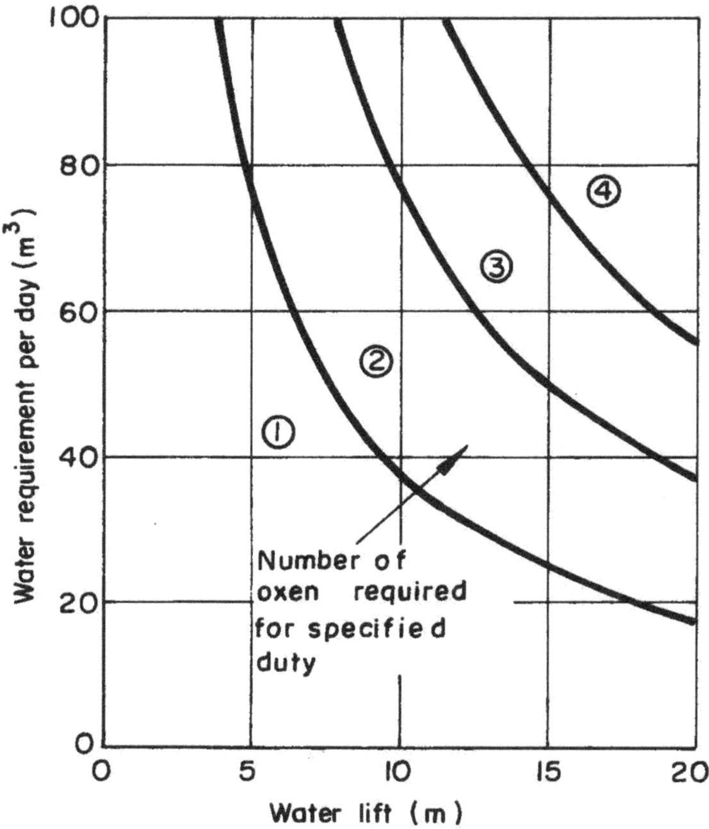


Figure 24. Number of oxen required to provide the specified quantity of water at different lifts. The curves have been calculated by assuming that an ox can provide 350 Watts of power for 5 hours per day, and that the efficiency of the water lifting device is 60%.

**Step 3. Size the Power Source and Pump**

A procedure for sizing solar pumps is described in Section 3.3.

For a windpump the required rotor size is determined by dividing the pumped water requirement, for the design month, by the pumped volume per m$^2$ of swept rotor area for the design month.

Figures 24 and 25 give an estimate of the number of animal or hand-pumps needed to provide the required quantity of water. Where actual measurements of the output of hand and animal pumps are known, these should be used in preference to the estimates of Figures 24 and 25.

Unless water requirements are extremely large a single diesel or kerosene pump will be sufficient. For example, the smallest sized diesel engine should provide an energy equivalent of 6000 m$^4$ per 12 hours of operation and a small kerosene engine may provide half this amount.

Steps 1 to 3 were covered in detail for a solar pump in Chapter 3.
Using the results obtained in Sections 3.1 to 3.3, the first part of Table 11 has been completed for the example solar pumping system.

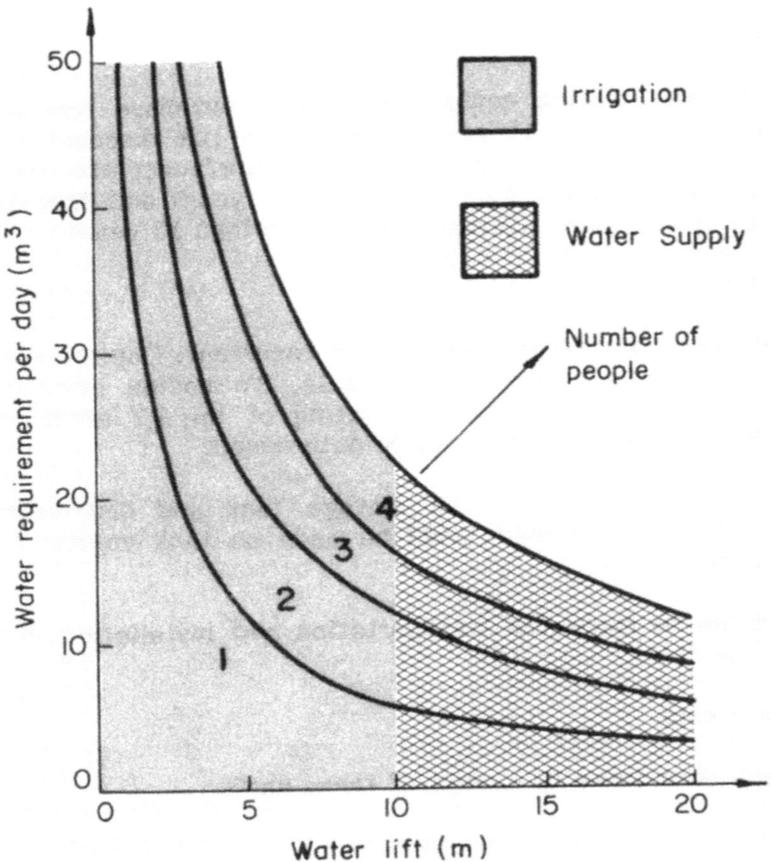


Figure 25. Number of people required to provide the specified quantity of water at different lifts. The curves have been derived by assuming that a single person can provide 60 watts of power for 4 hours per day and that the efficiency of the handpump is 60% (see example 1, Appendix 4).

For the final four steps the following data are required:

Economic:

- period of analysis.
- discount rate.
- inflation rates.

Costs for each component:.

- capital cost.
- annual maintenance cost.
- fuel, animal feeding or labour cost.

Technical:

- lifetime of each component in years.

The period of analysis should be at least equal to the economic lifetime of system, that is the time when the costs of maintenance make it less economical to use existing equipment than to purchase improved equipment. For the example, a 15 year period and a 10% discount rate have been chosen. It has been assumed that the motor/pump lifetime is 7.5 years, that the distribution pipe lifetime is 5 years and that the PV array and water source have lifetimes of greater than 15 years.

## Step 4. Capital Costs

Costs are divided into two types: capital and recurrent. Capital costs are dependent on component size. For example, PV module costs are about $8 per Wp (1985 prices). Once the rating of the PV array has been calculated the cost of the array can be determined.

To determine the capital cost of the storage tank and distribution systems (where included) a decision must be made on tank volume and pipe or channel sizes.

Due allowance should be made for transportation and installation when assessing the capital costs.

## Step 5. Recurrrent Costs

Recurrent costs are considered to consist of three parts:

Replacement costs occuring at intervals depending on the lifetime of each component. The number of replacements during the period of analysis is found, and the present worth of each replacement is then calculated by multiplying the initial capital cost of each component by the factor Pr. Pr is determined for the appropriate value of component inflation rate (i), discount rate (d) and year of replacement N) from Appendix 2. For most cases the component relative inflation rate will be zero. However, by allowing for an inflation rate, the relative cost of each component is allowed to change throughout the period of analysis. For example, it is likely that the present cost of PV modules will be reduced. Hence the replacement cost will be lower that the present cost and a negative inflation rate may be used to calculate the Present Worth.

Maintenance and repair costs occurring each year. The Present Worth of the maintenance cost for each component is found by multiplying the annual cost by the factor Pa as determined from Appendix 2.

Operating costs. These may be fuel costs, animal feeding costs or labour charges for operation and attendance. For diesel engines they are calculated on the basis of hours of operation. Figure 26 shows an estimate of the water pumped per hour for small 2.5 kW diesel engines with a fuel consumption of 1.5 litres per hour. From this graph the number of hours of operation needed to provide the required pumped volume can be determined. The annual fuel costs are then determined by multiplying the operating hours per year by the cost per hour.

The operating costs attributable to wages and animal feeding are obtained by multiplying the number of animals or humans (as determined in Step 3) by the daily charge rate. Attendance charges should also be included where necessary.

The PW of the operating costs are determined by multiplying the annual operating cost by the factor Pa from Appendix 2. If it is anticipated that the costs are going to increase, in real terms, an appropriate value for the inflation rate should be used.

### Step 6. Life Cycle Costs (LCC)

The life cycle costs are simply the sum of the Capital Cost and the Present Worth of the Recurrent Costs.

### Step 7. Unit Water Costs

The cost appraisal could end with the calculation of the life cycle costs. However it is more convenient to assess the pumping method by the cost of the water that the system provides. The unit water cost is determined by first converting the life cycle cost into an annual equivalent (annual equivalent life cycle cost or ALCC). Conversion of the LCC into the ALCC is the reverse process of discounting. The factor Pa must be calculated or obtained from Appendix 2, using the chosen discount rate (d) the period of analysis (N) and an inflation rate (i) of zero. The ALCC is obtained by dividing the LCC by the factor Pa.

The ALCC is then divided by the annual water requirement to give the unit water cost.

Steps 4 to 7 have been completed for the example system. The results are shown in Table 11. Since there is no storage tank, the capital cost comprises the water source cost (in this case \$40 to dig a 2 metre well), the PV array at \$8 per Wp for the modules, plus a miscellaneous cost of \$3.5 per Wp, a d.c motor/pump cost at \$2.00 per Wp and a distribution pipe cost at \$8 per metre of pipe, giving an installed capital cost of approximately \$7,570 for the complete system*.

During the 15 year period of analysis a replacement motor/pump is required at 7.5 years, and replacement pipes at years 5 and 10. The factor Pr for each replacement is obtained from Appendix 2.

For the motor/pump replacement with N=7.5 years, d=10% and i=0%, the factor Pr=0.49. The PW of the replacment cost is then 0.49 x 1080 = \$529. For the first pipe replacement, with N=5 years, d=10% and i=0%, Pr=0.62. For the second pipe replacement with N=10 years, d=10% and i=0%, the factor Pr =0.38. Hence the PW of the pipe replacements is (0.62 + 0.38) x 240 = \$240. (Note: it is assumed that the costs of these components do not change in real terms; an inflation rate (i) of zero is used).

---

* The solar pumping costs have been obtained using the values given in Section 4.3. For a system rated at 540 Wp the costs can be interpolated from Table 12. If projected costs were used, the capital cost for the complete pumping system would become \$4590; the PW of the recurrent costs would be \$1378 and the unit water cost would become \$6.3 cents per m $^3$.

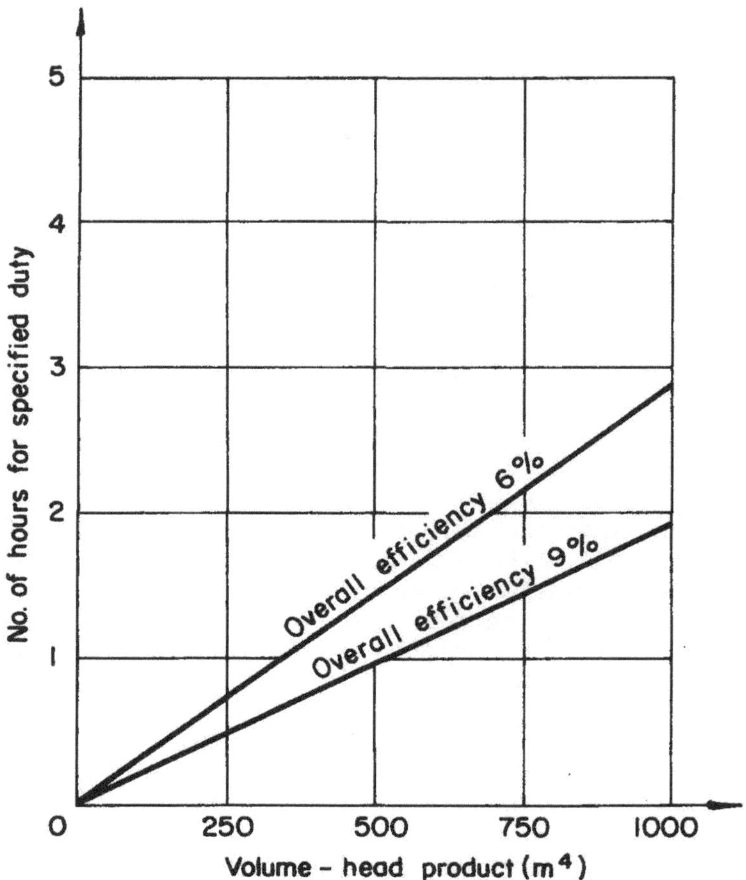


Figure 26. Number of hours required for a diesel pump to provide the specified water volume-head product. The volume-head product is obtained by multiplying the pumped volume in $m^3$ per day by the total system head in metres. The estimates are based on a diesel engine with a fuel consumption of 1.5 litres per hour and the specified efficiencies.

Annual maintenance costs of $80 are assumed. To find the PW of the maintenance costs the factor Pa is obtained from Appendix 2. For a period of analysis (N) = 15 years, (d) = 10% and (i) = 0%, the factor Pa = 7.61 giving a PW of $609 for the maintenance costs.

In this example, operating costs are assumed to be zero. The life cycle cost (LCC) is obtained by summing the capital cost with the present worth of the replacement and maintenance costs.

To obtain the unit water cost, Pa is obtained, from Appendix 2, for N = 15 years, d = 10% and i = 0%. Pa is equal to 7.61. The annual equivalent life cycle cost (ALCC) is the life cycle cost (LCC) divided by Pa, giving an ALCC of $8947 ÷ 7.61 = $1176. The annual water requirement is 12410 $m^3$, so the unit water cost is $1176 ÷ 12410 = 9.5 cents per $m^3$.

Two points should be noted in relation to this unit water cost:

> (i) The unit cost is the cost of the water provided to the crop and not the cost of water provided by the pump. (Since the field application efficiency is 60%, the actual annual volume of water pumped to meet the crop water requirement is $12410 \div 0.6 = 20683 \, m^3$.
>
> (ii) The unit cost is based on the cost of the complete system. It includes the costs of the water source and distribution network.


## 4.3. Guidance on Costs

In this section, typical costs and performance data have been analysed, taking account of experience obtained under the UNDP/World Bank Solar Water Pumping Project and on more recent work carried out by the Project Consultants. The information is included in order to provide the reader with a reference against which to compare the results of his/her estimates for a particular application.

### 4.3.1. Solar Pump Costs

Present and anticipated future PV solar pumping costs are given in Table 12. The main changes in future costs are expected to be in connection with PV module costs, for which three cases are considered:

(a) Present Costs

> i.e. 1985 costs with a module cost of $8 per Wp (fob point of manufacture);

(b) Projected Costs

> (estimated to be 1987) when due to the mass production of PV modules and dc motor/pump units, the costs will be reduced by 30%.

(c) Potential Costs

> (estimated to be 1993-1998) with motor/pump costs the same as for Target Costs but, due to changes in technology the PV module costs are assumed to have fallen to $2 per Wp (fob point of manufacture)

The capital costs in Table 12 are divided into three parts:

> (a) PV array. The cost is expressed in $ per peak electrical watt.

> (b) Motor/Pump. This cost is expressed in terms of $ per peak electrical watt. However, since the cost of a motor/pump unit is not directly proportional to hydraulic duty or electrical rating, costs are given for several ratings. At larger duties the cost per Wp is proportionally less. It should be noted that there are large variations in the costs of commercial motor/pumps and the

|  | Electrical Rating of Pump (Watts) | Present (1985) | Projected (1987) | Potential (1993-1998) |
|---|---|---|---|---|
| Module Cost ($/Wp) | All | 8 | 5 | 2 |
| Motor/Pump Cost (Fob) ($/Wp) | 200<br>400<br>600<br>1000 | 3.5<br>2.2<br>1.8<br>1.4 | 1.8<br>1.1<br>0.9<br>0.7 | 1.8<br>1.1<br>0.9<br>0.7 |
| Misc. Costs ($/Wp) | 200<br>400<br>600<br>1000 | 6.5<br>4.0<br>3.2<br>2.5 | 5.7<br>3.9<br>3.2<br>2.8 | 5.7<br>3.9<br>3.2<br>2.8 |
| Total Installed Solar Pump Cost ($/Wp) | 200<br>400<br>600<br>1000 | 18<br>14<br>13<br>12 | 13<br>10<br>9<br>8 | 10<br>7<br>6<br>5 |

Table 12   Present and Projected Solar Pump Costs

values given are only averages; for example, more efficient motor/ pumps could be more expensive.

(c) Miscellaneous costs, covering power conditioning, pipework, foundations, shipping, transport and labour, are expressed in $ per Wp for several electrical ratings since they become proportionally less for large systems. It is anticipated that these costs will also be reduced as power conditioning equipment becomes cheaper.

The resulting total installed solar pumping system cost is dependent on the system size. The cost is expressed in $ per peak electrical watt (i.e. PV array rating). The costs per peak electrical watt are, of course, dependent on the efficiency of the subsystem, since this relates the hydraulic power to the electrical power. The system costs, shown in Table 12, have been calculated assuming a peak subsystem power efficiency of 60%. At present, with this subsystem peak power efficiency, typical system costs are $12 to $18 per installed peak electrical watt. Within the next two years it is anticipated that this cost will be reduced to $8 to $13 per peak electrical watt and by the year 2000, if new solar PV technologies emerge, the installed cost could be between $5 and $10 per peak electrical watt.

Recurrent costs for solar pumps are difficult to estimate due to the limited operating experience. Table 13 gives the values of lifetime and maintenance costs that were assumed for the most recent work on the UNDP/World Bank Small Scale Solar Pumping Project. Recurrent costs for motor/pump units were calculated from the number of operating hours. For a system with a constant daily water demand it is usual to assume a typical number of operating hours per year of approximately 2500, giving a motor/pump lifetime of about 8 years.

| | Recurrent Maintenance Cost | | Lifetime | |
|---|---|---|---|---|
| | $ per year | $ per 1000 operating hours | years | operating hours |
| PV Array | 50 | - | 15 | - |
| Motor/Pump | - | 12 | - | 20,000 |

Table 13.    Data used to Calculate Recurrent Costs for UNDP Project GLO/80/003 "Small Scale Solar Pumping Systems"

### 4.3.2.  Other Pumping System Costs

Table 14 gives typical capital and recurrent costs for other pumping methods. These costs were based on a variety of sources ranging from surveys carried out in Bangladesh, Kenya, Thailand and the United Kingdom to a worldwide survey of windpump costs.*

The following points should be noted in relation to these costs:

o   Reliable operating and maintenance cost data are extremely limited for all methods of pumping. In particular there are few data on diesel engines; consequently a range of costs is presented.

o There is a large range of windpump costs from $600 per $m^2$ for the windpumps manufactured in developed countries to $50 $m^2$ for locally manufactured designs.

o   Operating costs for handpumps depend on the cost attributed to local labour. For the UNDP/World Bank Small Scale Solar Pumping

---

* Windpump costs are reported in "Wind Technology Assessment Study" by I T Power Limited.   Published by The World Bank, 1818 H Street NW, Washington DC 20433, USA.

| Pump | Wind | Diesel | Kerosene | Animal | Hand |
|---|---|---|---|---|---|
| Hydraulic Rating | dependent on wind regime | 1.3 kW | 550W | 210W | 36W |
| Capital Cost | $300 per m$^2$ of rotor (1m<dia<10m) | $2125 for a 2.5 kW engine = $1.42 per W of hydraulic power | $440 for a 1.1kW engine = $0.8 per W of hydraulic power | $250 per animal = $5.2 per W of hydraulic power | $216-290 per handpump = $6 - $8 per W of hydraulic power |
| Maintainance Cost<br><br>$ per year<br><br>and/or<br><br>$ per 1000 operating hours | 50<br><br><br><br>6 | -<br><br><br><br>200-400 | -<br><br><br><br>200-400 | 10<br><br><br><br>- | 50<br><br><br><br>- |
| Operating cost | 0 | 40 to 80 cents per litre | 40 to 80 cents per litre | $2.25 per animal day | $1 per man day |

Table 14. Data Used to Calculate Capital and Recurrent Costs for UNDP/World Bank Project GLO/80/003 "Small Scale Solar Pumping Systems"

Project a value of $1 per man-day was attributed to labour for irrigation pumping and zero cost was assumed for labour on village water supplies.

### 4.3.3 Unit Water Costs

An idea of the current and future unit water costs of solar pumps, when compared to diesel and windpumps, can be obtained from Figures 27 and 28. These figures have been prepared using the cost and lifetime data of Sections 4.3.1 and 4.3.2 with the cost comparison methodology described in Section 4.2. The life cycle costs have been determined for a 30 year period at a 10% discount rate. (This choice of the period of analysis was used because it allows for replacements of all major system components. Shorter periods could be used but would have little effect on the end result.)

In these Figures the unit water cost due to the pump (i.e. excluding water source, storage and distribution) is expressed in $ per m$^4$ and is shown as a function of the annual average energy equivalent in m$^4$ per day. The energy equivalent is calculated by taking the product of annual crop, village or livestock water requirement (m$^3$/year) and the total system head, and dividing by 365 days. The cost in $ per m$^4$ must be multiplied by the system head to obtain the cost per cubic metre of water.

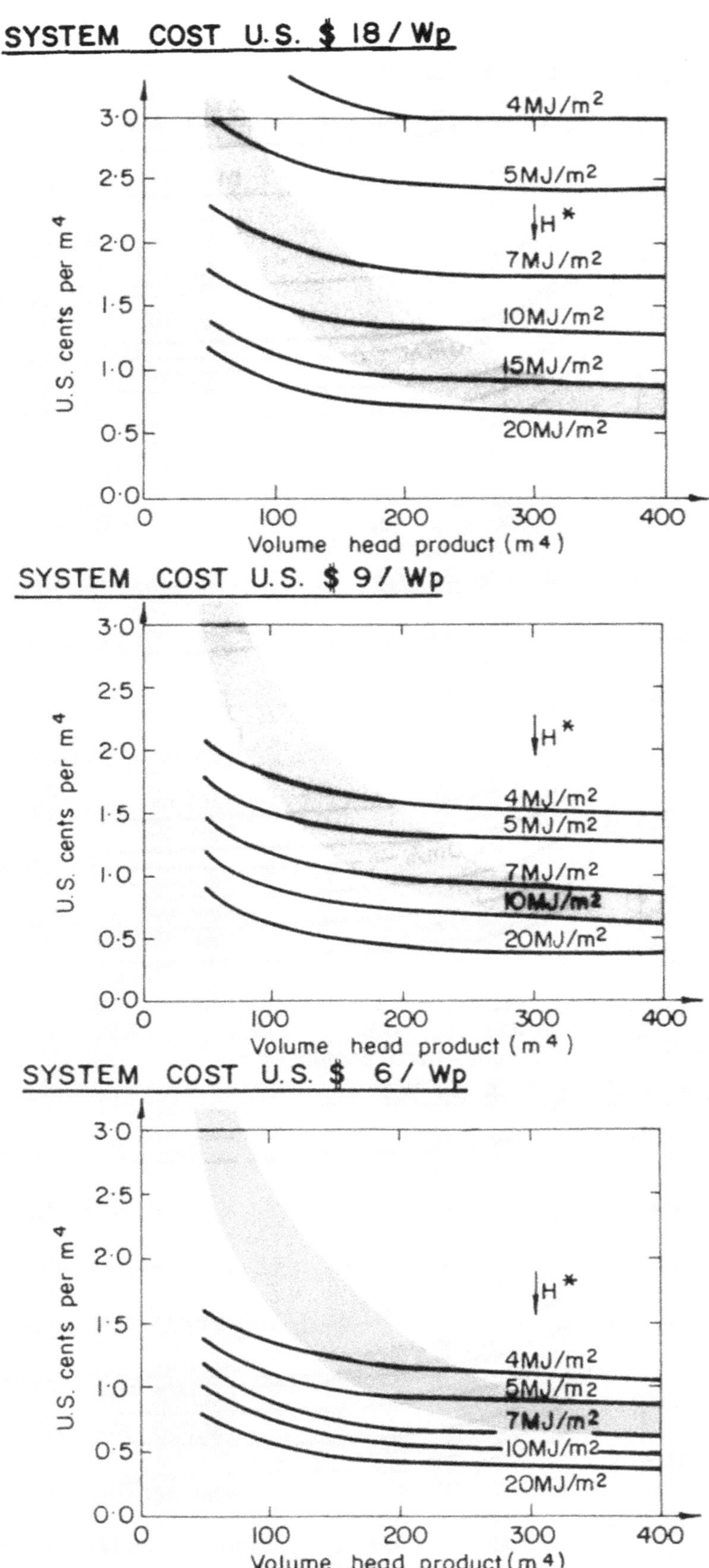


Figure 27. Comparison of unit water costs between solar and diesel pumps. The shaded area represents a typical range of diesel pumping costs.

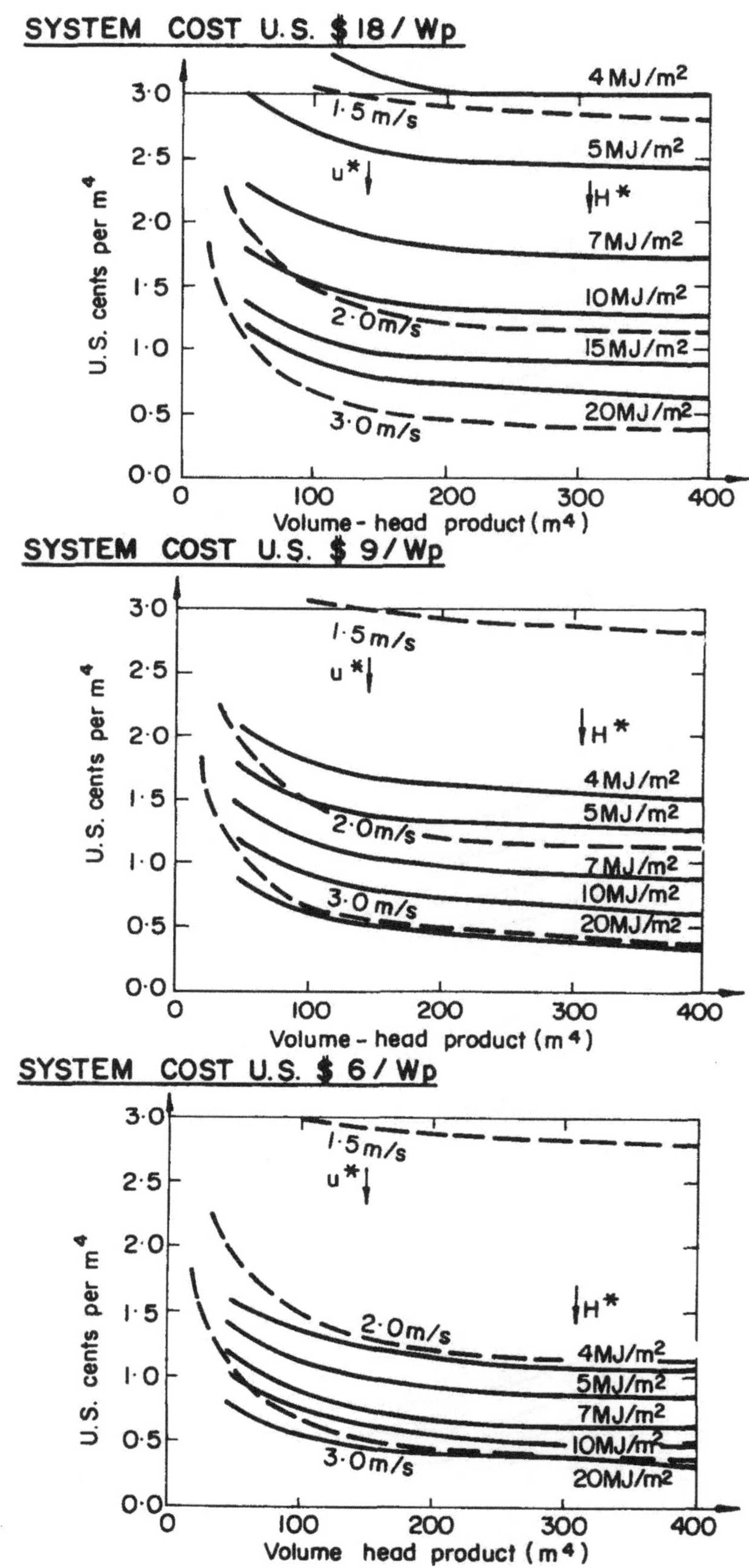



Figure 28. Comparison of unit water costs between solar and windpumps

68

For solar and windpumps the unit water cost is dependent on the following location dependent parameters:

$H^*$ = design month solar irradiation $(MJ/m^2)$ x
(storage and distribution efficiency)
÷ (design month demand factor)

$u^*$ = design month mean wind speed (m/s) x
(storage and distribution efficiency)$^{\frac{1}{3}}$
÷ (design month demand factor)$^{\frac{1}{3}}$

The design months are determined using the procedure outlined in Section 4.2, and the design month demand factor is defined as the ratio of the design month water requirement (in $m^3$ per day) to the annual average water requirement (in $m^3$ per day).

For the diesel pump the unit water cost is given for a range of conditions representing the range of operating and maintenance costs given in Table 14.

Figures 27 and 28 illustrate several characteristics of solar pump economic viability.

o   There is a marked dependence of water costs on the values of $H^*$ and $u^*$, which depend on water demand pattern, efficiency of distribution and meteorological conditions.

o   the effect of $H^*$ will become less if future cost reductions are achieved.

o   at present, solar pumps provide cheaper water than diesel pumps for low water requirements (generally low head applications), in locations with high solar irradiation and little variation in month by month water demand (e.g. Vh < 200 $m^4$ for $H^*$ = 20 $MJ/m^2$). Such circumstances are not common. However, other factors such as reliability and ease of maintenance can be expected to encourage the use of solar pumps in conditions where they are not the cheapest option.

o   windpumps are and will probably remain cheaper than solar or diesel in locations having values of $u^*$ greater than 3 m/s

o if solar pump costs of $6 per Wp are reached, solar pumps can be expected to become cost competitive with diesel pumps for most small scale applications in locations where $H^*$ is greater than 7 $MJ/m^2$.

**Example of a cost appraisal using Figures 27 and 28**

What are the unit water costs for a) a solar pump, b) a windpump and c) a diesel pump to provide the water requirements for the irrigation system used as an example in Chapter 3? Compare a solar pump costing \$18/Wp with one costing \$9/Wp. (Assume that the wind design month is February with a mean wind speed estimated to be between 2.0 and 2.5 m/s.)

1. The average daily crop water requirement is $12410 \div 365 = 34$ m$^3$. Assuming an average system head of **2.2 m**, the energy equivalent is $34 \times 2.2 = 74.8$m$^4$.

2. For the solar pump the design month is May with a solar irradiation of 16 MJ/m$^2$. The crop water requirement in May is 60 m$^3$ per day. Hence the demand factor for this month is $60 \div 34 = 1.76$. The distribution efficiency is 60%, giving $H^* = 16 \times 0.6 \div 1.76 = 5.45$ MJ/m$^2$. From Figure 27, the unit water cost at a capital cost of \$18 per Wp for an energy equivalent of 74.8 m$^4$ and a $H^*$ value of 5.45 MJ/m$^2$ is 2.8 cents per m$^4$. At a head of 2.2 m this is equivalent to $2.2 \times 2.8 = 6.2$ cents per m$^3$.

From Figure 27, at a capital cost of \$9 per Wp the unit water cost will be 1.6 cents per m$^4$ equivalent to $(1.6 \times 2.2) = 3.5$ cents per m$^3$.

3. For the windpump the design month is February with a crop water requirement of 50 m$^3$ per day. The demand factor for February is $50 \div 34 = 1.47$ giving $u^*$ values in the range from $2.0 \times (0.6 \div 1.46)^{\frac{1}{3}} = 1.5$ m/s to $2.5 \times (0.6 \div 1.46)^{\frac{1}{3}} = 1.9$ m/s. From Figure 28 the unit water cost lies in the range from 3.2 to 2.0 cents per m$^4$, equivalent to water volume costs in the range $3.2 \times 2.2 = 7.0$ to $2.0 \times 2.2 = 4.4$ cents per m$^3$.

4. For the diesel pump the unit water costs for an energy equivalent of 74.8 m$^4$ range from 2.1 to 3.2 per m$^4$, equal to **4.6 to 7.0** cents per m$^3$.

Figure 29 shows the results obtained for this example.

The conclusion from this example is that, for the low head and other favourable conditions assumed, solar pumping costs are presently well within the range of the likely costs of the diesel and wind pumping alternatives. In these circumstances, the greater reliability of solar pumps owing to their freedom from external fuel supplies and lower maintenance requirements may well tilt the decision in their favour, especially in comparison with diesel. The example also shows that when projected solar pumping costs are achieved, (\$8–13 per Wp) solar will clearly be the least cost option under the assumed conditions.

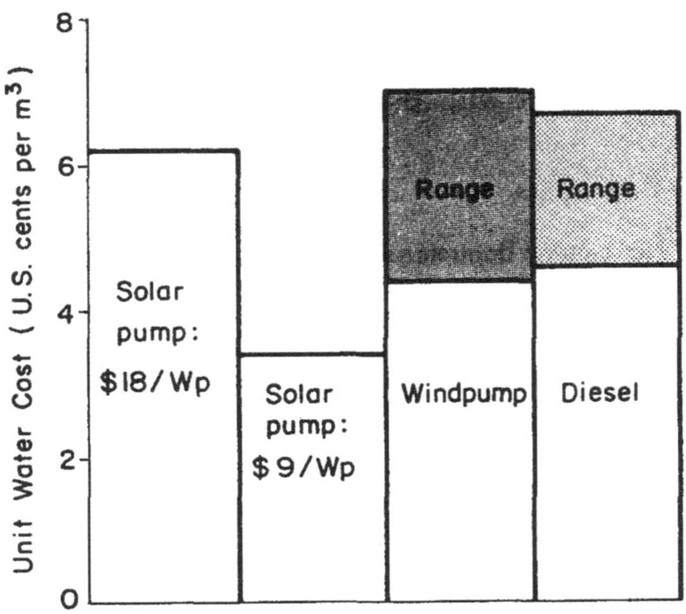



Figure 29. Histogram showing unit water costs for the example system.

# 5. PROCUREMENT, INSTALLATION AND MONITORING

## 5.1. Procurement

Five stages for the specification and procurement of a solar PV pump are recommended, as described briefly below.

**(1)   Assess Solar Pumping Viability and Estimate Costs**

Before contacting suppliers, an initial appraisal of solar pumping should be made in accordance with the guidelines given in Chapters 1 and 3.

**(2)   Prepare Tender Documents**

A suggested format for tender documents is given in Appendix 6.

**(3)   Issue a call for Tenders**

Letters may be sent to suppliers with a brief description of the required system. Interested suppliers will then reply with a request for the tender documents.

**(4)   Preliminary Evaluation**

Each tender should be checked to ensure that:

> (a) the system being offered is complete, and includes spare parts and installation and operating instructions.

> (b) the system being offered can be delivered within the maximum period specified, and

> (c) that an appropriate warranty can be provided.

**(5)   Detailed Assessment**

A detailed assessment of each tender should be made under the following four headings with approximately equal importance ascribed to each heading.

> (a) <u>Compliance with specification</u>

> The output of the system should be assessed taking into account any deviations from the specification proposed by the tenderer.

> (b) <u>System design</u>

> The suitability of the equipment for the intended use should be assessed taking into account operation and maintenance requirements, general complexity, safety features etc.

> The equipment life should be assessed with regard to bearing

brushes and other parts liable to wear and tear.

The content of the information supplied to support the tender should be assessed, in particular the provision of general assembly drawings and performance information.

(c) Capital cost

Capital costs should be compared, allowing for any deviations from the specification proposed by the tenderer. This can be done by comparing the capital cost per $m^4$ of water delivered by the system.

(d) Overall credibility of tender.

The experience and resources of the tenderer relevant to solar pumping technology in developing countries should be assessed together with the tenderer's ability to provide a repair and spare parts service should problems be experienced with the solar pump. A reasonable warranty, at least relating to spare parts, should be provided.

## 5.2. Installation and Operation

### 5.2.1. Installation and Commissioning

Arrays

Arrays should be sited in an unshaded position. Sites should be checked for trees and buildings that could shade the array, remembering that the sun's position changes with season, and that trees grow. The arrays should be away from possible damage by floods, and fenced in to protect them from vehicles, children, animals etc. Portable arrays should be temporarily fixed down, or weights should be used to stop them from blowing over.

Erection should be carried out using the manufacturer's installation instructions. Care should be taken not to lose nuts, bolts and washers. All connections should be cleaned on assembly and checked for tightness after erection. Electrical cables should be protected from possible damage and an earth (ground) connection installed where necessary.

Floating Motor/Pump Units

Seals, cable glands and the flotation device fixings should be checked as appropriate. The unit should not be lifted or supported by its cable - a rope should be attached to a suitable point to lower the unit onto the water. Discharge hoses may also need support in wells or fast flowing rivers. Sufficient cable and hose should be allowed to accommodate changes in the water level.

<u>Surface Mounted Suction Pumps</u>

The suction lift should be as small as practical. Suction pipes must be leak free and the pipe needs to rise steadily so that air is not trapped. The priming chamber should be filled before the pump is operated, and checks should be made to ensure that the inlet to the suction pipe is free from sediment, floating debris etc, and that the direction of rotation of the motor is correct and that the foot valve (where fitted) is operating correctly.

<u>Surface Mounted Motor with Submerged Pump</u>

A skilled fitter should assemble the vertical drive shaft. The tightness and adjustment of clearances is vital. The rising main must be primed before starting, and the direction of rotation and well level controls (if fitted) should be checked. The pump and shaft must not run dry.

<u>Submersible Motor/Pump Units</u>

All cable connections and seals should be checked for water tightness as far as possible, and a supporting wire should be attached to the motor/pump unit if plastic riser pipes are used. Start up instructions should be followed closely. Normally the direction of rotation is checked by comparing flow and pressure with the electrical connections the correct way and with them reversed. However in some cases reversing the connections could damage the unit. The manufacturer's instructions should be followed. Centrifugal pumps will still pump water when running backwards, but less efficiently.

<u>Jack and Piston Pumps</u>

When jack and piston pumps are mounted on an open well, the well must be covered while working above it. Ensure that all rod and riser pipe joints are tight. A coarse strainer is often used on the pump inlet, but filters should be fitted on the discharge side of the pump where they can be cleaned. Any balance weights, pulleys etc. should be adjusted in accordance with the manufacturer's instructions. The pump should be fitted with guards or fenced to keep animals and people away from moving parts.

### 5.2.2. Maintenance

Solar pumps should not normally require more than simple maintenance functions which only demand rather basic skills. The main problem with them at present is lack of familiarity; the "black box" nature of solar pump components makes their function appear mysterious and may discourage farmers or local mechanics from trying to correct any faults which develop.

The only relatively frequent maintenance function necesary with most systems is to clean the photovoltaic array from time to time if it gets covered with dust. The motor-pump subsystem is not dissimilar to a mains electric pump in terms of maintenance requirements, which in common with mains electrical systems should be minimal.

Possibly a primary cause of failures results from damage caused by animals or people to the relatively fragile photovoltaic array. It is generally necessary to install solar pumps within a fenced enclosure for protection. The fence therefore needs to be kept in good condition and the gate should be safely secured.

There are a number of faults that can arise which can readily be corrected by the user without special tools or equipment; for example:

o  poor electrical connection caused by dirty, wet or corroded terminals or plugs.

o  blocked strainers and filters on the pump

o  failure of suction pump due to loss of prime caused by faulty foot-valve or air leaks in suction line.

o  leaking pipe or hose connections

o  leaking pump gland seal

o  some motors need replacement brushes; this is usually a simple operation described in the handbook, (far simpler than for example servicing a small engine powered pump)

o  where gearboxes or mechanical transmission are involved (usually with positive displacement pumps only) then occasional oil changes or greasing may be necessary and/or belts or chains may need occasional retensioning or adjustment.

When a vital system component suffers damage, which should be a rare occurence, then it is possible that a skilled engineer will be required to identify the cause of failure. It would generally be unusual to dismantle and attempt to repair a failed component in the field; normally the suspected component would be removed and a replacement subsituted. Dismantling and re-assembly is generally a simple matter requiring no more than a few spanners and a screwdriver, so providing the failed item can be identifieid and a replacement obtained, it is not difficult to make the substitution.


## 5.3. Monitoring and Evaluation

### 5.3.1.  The Need for Monitoring

Photovoltaic solar pumps are a relatively new technology, so it is essential to obtain data on their performance to build up information on their long term technical, economic and social viability. The objectives of a monitoring programme will usually include the following aspects:

Methods of application: to gain first hand experience of the ways in which the solar pumps are used by farmers and villagers;

Economics: to gain information on the costs and benefits of PV systems;

User reaction: to find ways in which the systems and their methods of application can be amended to make them more acceptable;

<u>Performance</u> <u>and</u> <u>reliability</u>: to increase the data base on system performance and, to obtain data on component lifetimes and maintenance requirements.


## 5.3.2. Methods of Evaluation

The simplest method of performance evaluation is to take daily readings of:

o    global solar irradiation in the plane of the PV array;
o    volume of water pumped;
o    static head.

This will enable the hydraulic energy and hence system efficiency to be obtained for different values of daily solar irradiation. A format sheet for recording such measurements is shown in Table 15.

As indicated in Chapters 2 and 3 the most important parameter for system sizing and economic evaluation is the daily subsystem efficiency. This can be obtained from the above readings by assuming an average daily efficiency for the PV array. However, a more thorough approach would include measurements of the array energy output. Scatter diagrams can then be obtained of subsystem efficiency against daily solar irradiation as shown in Figure 30.

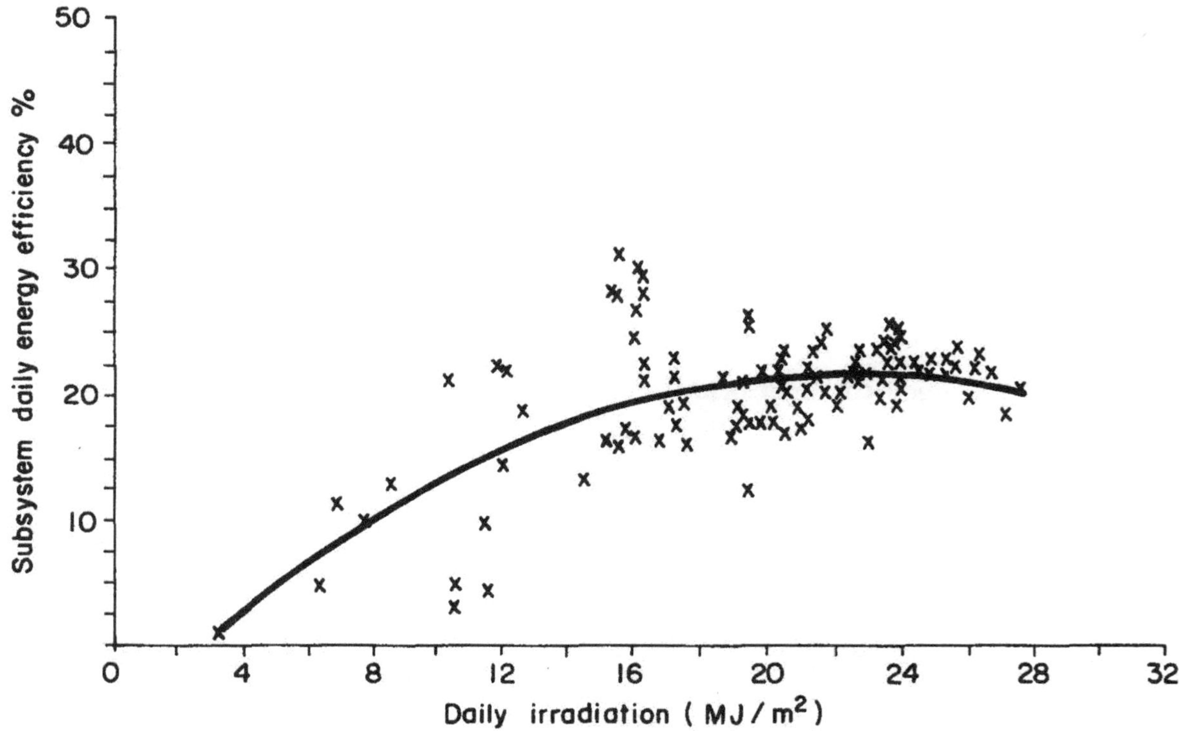

Figure 30. Example showing the variation of subsystem efficiency with solar irradiation. Such data may be obtained from field trials. They are important for system sizing and economic evaluation. (The results shown here were obtained from field trials in Mali as part of Phase I of the UNDP/World Bank Solar Pumping Project).

| SOLAR PUMP PERFORMANCE | | | | | | |
|---|---|---|---|---|---|---|
| Location .....................    Month ............... | | | | | | |
| System Type .......................................... | | | | | | |
| .......................................... | | | | | | |
| Day | Time | Solarimeter Reading | Flow Meter Reading | Static Head | Solar Irradiation $MJ/m^2/day$ | Volume Pumped $m^3/day$ | Initials of Recorder |
| 1 | | | | | | | |
| 2 | | | | | | | |
| 3 | | | | | | | |
| 4 | | | | | | | |
| 5 | | | | | | | |
| 6 | | | | | | | |
| 7 | | | | | | | |
| 8 | | | | | | | |
| 9 | | | | | | | |
| 10 | | | | | | | |
| 11 | | | | | | | |
| 12 | | | | | | | |
| 13 | | | | | | | |
| 14 | | | | | | | |
| 15 | | | | | | | |
| 16 | | | | | | | |
| 17 | | | | | | | |
| 18 | | | | | | | |
| 19 | | | | | | | |
| 20 | | | | | | | |
| 21 | | | | | | | |
| 22 | | | | | | | |
| 23 | | | | | | | |
| 24 | | | | | | | |
| 25 | | | | | | | |
| 26 | | | | | | | |
| 27 | | | | | | | |
| 28 | | | | | | | |
| 29 | | | | | | | |
| 30 | | | | | | | |
| 31 | | | | | | | |

Table 15  Format sheet for Recording Solar Pump Performance

Instantaneous measurements of power can also be made, but these are of more use for system design improvements than for economic evaluations.

In addition to taking performance measurements, a log book should be used to record problems, breakdowns and maintenance requirements.

### 5.3.3. Instrumentation

Instrumentation selected for performance monitoring should meet the following requirements:

- o    suitable for field use;
- o    allow on site assessment of performance;
- o    have suitable accuracy (better than 5%);
- o    require only battery power and have a long battery life before discharge, (say greater than 1 month)

Three types of instrument are required in order to complete the daily recordings outlined above:

> (a) a solarimeter to measure solar irradiance. This can be either an accurate thermopile device or a less accurate but cheaper photovoltaic device. The solarimeter requires an integrating meter to record the daily solar irradiation.

> (b) a flow meter to record flow rate. The head loss caused by the flow meter should be kept to a minimum and the device should be capable of measuring flows in the range expected from the pump. The flow meter also requires an integrating meter, so that the volume of water pumped in a day can be obtained.

> (c) a well dipper with a water sensitive transducer that permits measurements of water level to be obtained easily. If the water source is a borehole, sufficient access may need to be made at the well head for measurements of water level. The well dipping is usually performed manually and a reading is recorded daily.

If measurements of subsystem efficiency are to be made, an energy meter will also be required to measure the electrical output of the array.

<u>**APPENDIX 1  SOLAR RADIATION DATA**</u>

The average daily global irradiation for a month can be obtained by multiplying the global irradiation outside the atmosphere, obtained from Table A1, by the clearness index obtained from Figures A1 to A12. The clearness indices are accurate to $\pm$ 10%, hence the resulting solar radiation will be accurate to $\pm$ 10% since the extra-terrestrial radiation is a fixed property of the sun.

To calculate the global irradiation on a tilted surface multiply the horizontal irradiation by the tilt factors given in Tables A2 to A14.  In these tables the surface inclination is specified relative to a horizontal surface.  The tilt factors have been calculated using geometrical formulae relating energy on a tilted surface to energy on a horizontal surface. They can be taken to be accurate to 2 to 3 percent.

Extra-terrestrial Global Irradiation (MJ/m$^2$ per day)

Month for Northern Hemisphere

| Latitude Degrees | Jan | Feb | Mar | Apr | May | Jun | Jul | Aug | Sept | Oct | Nov | Dec | Latitude Degrees |
|---|---|---|---|---|---|---|---|---|---|---|---|---|---|
| 0 | 36 | 37 | 38 | 36 | 34 | 33 | 34 | 35 | 37 | 37 | 36 | 35 | 0 |
| 5 | 34 | 36 | 37 | 37 | 36 | 35 | 35 | 36 | 37 | 36 | 34 | 33 | 5 |
| 10 | 32 | 34 | 37 | 38 | 37 | 37 | 37 | 37 | 37 | 35 | 32 | 31 | 10 |
| 15 | 29 | 32 | 36 | 38 | 38 | 38 | 38 | 38 | 36 | 33 | 30 | 28 | 15 |
| 20 | 27 | 30 | 34 | 38 | 39 | 39 | 39 | 38 | 35 | 31 | 27 | 25 | 20 |
| 25 | 24 | 28 | 33 | 37 | 39 | 40 | 40 | 38 | 34 | 29 | 25 | 23 | 25 |
| 30 | 21 | 26 | 31 | 37 | 40 | 41 | 40 | 37 | 33 | 27 | 22 | 20 | 30 |
| 35 | 18 | 23 | 29 | 36 | 40 | 41 | 40 | 37 | 31 | 25 | 19 | 17 | 35 |
| 40 | 15 | 20 | 27 | 34 | 39 | 41 | 40 | 36 | 29 | 22 | 16 | 14 | 40 |
| 45 | 12 | 18 | 25 | 33 | 39 | 41 | 40 | 35 | 27 | 19 | 13 | 11 | 45 |
| 50 | 9 | 15 | 22 | 31 | 38 | 41 | 40 | 34 | 25 | 16 | 10 | 8 | 50 |
| 55 | 6 | 12 | 20 | 29 | 37 | 41 | 39 | 32 | 23 | 14 | 7 | 5 | 55 |
| 60 | 3 | 9 | 17 | 27 | 36 | 41 | 38 | 30 | 20 | 11 | 4 | 2 | 60 |

Jul  Aug  Sep  Oct  Nov  Dec  Jan  Feb  Mar  Apr  May  Jun

Month for Southern Hemisphere

Table A1.  Average Daily Global Irradiation for a horizontal surface outside the earth's atmosphere.

| Clearness Index | Tilt (degrees) | Jan | Feb | Mar | Apr | May | Jun | Jul | Aug | Sep | Oct | Nov | Dec |
|---|---|---|---|---|---|---|---|---|---|---|---|---|---|
| | 10 | 1.03 | 1.01 | 1.00 | 0.97 | 0.96 | 0.95 | 0.95 | 0.97 | 0.99 | 1.01 | 1.03 | 1.04 |
| | 20 | 1.04 | 1.01 | 0.97 | 0.93 | 0.90 | 0.88 | 0.89 | 0.92 | 0.96 | 1.00 | 1.04 | 1.05 |
| 0.3 | 30 | 1.04 | 0.99 | 0.93 | 0.87 | 0.82 | 0.80 | 0.81 | 0.85 | 0.91 | 0.97 | 1.03 | 1.05 |
| | 40 | 1.01 | 0.95 | 0.88 | 0.80 | 0.74 | 0.71 | 0.73 | 0.78 | 0.85 | 0.93 | 1.00 | 1.03 |
| | 50 | 0.96 | 0.89 | 0.81 | 0.71 | 0.65 | 0.62 | 0.64 | 0.69 | 0.77 | 0.87 | 0.95 | 0.99 |
| | 60 | 0.90 | 0.82 | 0.72 | 0.62 | 0.56 | 0.54 | 0.55 | 0.60 | 0.69 | 0.80 | 0.89 | 0.93 |
| | 10 | 1.05 | 1.02 | 1.00 | 0.96 | 0.94 | 0.93 | 0.93 | 0.95 | 0.98 | 1.02 | 1.04 | 1.06 |
| | 20 | 1.07 | 1.03 | 0.97 | 0.91 | 0.86 | 0.84 | 0.85 | 0.89 | 0.95 | 1.01 | 1.06 | 1.09 |
| 0.4 | 30 | 1.08 | 1.01 | 0.93 | 0.84 | 0.77 | 0.73 | 0.75 | 0.81 | 0.89 | 0.99 | 1.06 | 1.10 |
| | 40 | 1.06 | 0.97 | 0.86 | 0.75 | 0.66 | 0.63 | 0.64 | 0.72 | 0.82 | 0.94 | 1.04 | 1.08 |
| | 50 | 1.01 | 0.91 | 0.78 | 0.65 | 0.55 | 0.51 | 0.53 | 0.61 | 0.73 | 0.87 | 0.99 | 1.05 |
| | 60 | 0.95 | 0.83 | 0.69 | 0.54 | 0.45 | 0.41 | 0.43 | 0.50 | 0.63 | 0.79 | 0.92 | 0.99 |
| | 10 | 1.06 | 1.03 | 1.00 | 0.95 | 0.92 | 0.90 | 0.91 | 0.94 | 0.98 | 1.02 | 1.06 | 1.07 |
| | 20 | 1.10 | 1.04 | 0.97 | 0.89 | 0.82 | 0.79 | 0.81 | 0.86 | 0.94 | 1.02 | 1.09 | 1.12 |
| 0.5 | 30 | 1.12 | 1.03 | 0.92 | 0.80 | 0.71 | 0.67 | 0.69 | 0.77 | 0.88 | 1.00 | 1.10 | 1.14 |
| | 40 | 1.10 | 0.99 | 0.85 | 0.70 | 0.59 | 0.54 | 0.56 | 0.66 | 0.80 | 0.95 | 1.08 | 1.14 |
| | 50 | 1.06 | 0.93 | 0.76 | 0.59 | 0.46 | 0.41 | 0.44 | 0.54 | 0.70 | 0.88 | 1.03 | 1.10 |
| | 60 | 1.00 | 0.84 | 0.65 | 0.46 | 0.34 | 0.30 | 0.32 | 0.41 | 0.58 | 0.79 | 0.96 | 1.04 |
| | 10 | 1.08 | 1.04 | 1.00 | 0.95 | 0.90 | 0.88 | 0.89 | 0.93 | 0.98 | 1.03 | 1.07 | 1.09 |
| | 20 | 1.14 | 1.06 | 0.97 | 0.87 | 0.79 | 0.75 | 0.77 | 0.84 | 0.93 | 1.03 | 1.12 | 1.16 |
| 0.6 | 30 | 1.16 | 1.05 | 0.91 | 0.77 | 0.65 | 0.60 | 0.63 | 0.72 | 0.86 | 1.01 | 1.13 | 1.19 |
| | 40 | 1.15 | 1.01 | 0.83 | 0.65 | 0.51 | 0.45 | 0.48 | 0.60 | 0.77 | 0.96 | 1.12 | 1.19 |
| | 50 | 1.11 | 0.94 | 0.74 | 0.52 | 0.37 | 0.30 | 0.33 | 0.46 | 0.66 | 0.89 | 1.08 | 1.16 |
| | 60 | 1.04 | 0.85 | 0.62 | 0.38 | 0.23 | 0.17 | 0.20 | 0.32 | 0.53 | 0.79 | 1.00 | 1.10 |
| | 10 | 1.10 | 1.06 | 1.00 | 0.93 | 0.88 | 0.86 | 0.87 | 0.91 | 0.98 | 1.04 | 1.09 | 1.12 |
| | 20 | 1.17 | 1.08 | 0.97 | 0.84 | 0.74 | 0.70 | 0.72 | 0.80 | 0.92 | 1.05 | 1.15 | 1.20 |
| 0.7 | 30 | 1.21 | 1.07 | 0.91 | 0.73 | 0.59 | 0.52 | 0.56 | 0.67 | 0.84 | 1.03 | 1.18 | 1.25 |
| | 40 | 1.21 | 1.03 | 0.82 | 0.59 | 0.42 | 0.35 | 0.38 | 0.53 | 0.74 | 0.97 | 1.17 | 1.26 |
| | 50 | 1.17 | 0.96 | 0.71 | 0.44 | 0.25 | 0.17 | 0.21 | 0.37 | 0.61 | 0.89 | 1.13 | 1.23 |
| | 60 | 1.1 | 0.87 | 0.58 | 0.28 | 0.10 | 0.03 | 0.06 | 0.21 | 0.47 | 0.79 | 1.05 | 1.17 |

Table A2  Tilt factors for latitude 0 degrees

<table>
<tr><th>Clearness Index</th><th>Tilt (degrees)</th><th colspan="12">Month for Northern Hemisphere</th></tr>
<tr><th></th><th></th><th>Jan</th><th>Feb</th><th>Mar</th><th>Apr</th><th>May</th><th>Jun</th><th>Jul</th><th>Aug</th><th>Sep</th><th>Oct</th><th>Nov</th><th>Dec</th></tr>
<tr><td rowspan="6">0.3</td><td>10</td><td>1.04</td><td>1.02</td><td>1.00</td><td>0.98</td><td>0.96</td><td>0.95</td><td>0.96</td><td>0.97</td><td>0.99</td><td>1.02</td><td>1.03</td><td>1.04</td></tr>
<tr><td>20</td><td>1.06</td><td>1.02</td><td>0.98</td><td>0.94</td><td>0.91</td><td>0.89</td><td>0.90</td><td>0.93</td><td>0.97</td><td>1.01</td><td>1.05</td><td>1.07</td></tr>
<tr><td>30</td><td>1.06</td><td>1.01</td><td>0.95</td><td>0.89</td><td>0.84</td><td>0.82</td><td>0.83</td><td>0.87</td><td>0.93</td><td>0.99</td><td>1.05</td><td>1.07</td></tr>
<tr><td>40</td><td>1.04</td><td>0.97</td><td>0.90</td><td>0.82</td><td>0.76</td><td>0.74</td><td>0.75</td><td>0.80</td><td>0.87</td><td>0.95</td><td>1.02</td><td>1.06</td></tr>
<tr><td>50</td><td>1.00</td><td>0.92</td><td>0.83</td><td>0.74</td><td>0.68</td><td>0.65</td><td>0.66</td><td>0.72</td><td>0.80</td><td>0.90</td><td>0.98</td><td>1.02</td></tr>
<tr><td>60</td><td>0.94</td><td>0.85</td><td>0.75</td><td>0.65</td><td>0.59</td><td>0.56</td><td>0.57</td><td>0.63</td><td>0.72</td><td>0.83</td><td>0.92</td><td>0.97</td></tr>
<tr><td rowspan="6">0.4</td><td>10</td><td>1.06</td><td>1.03</td><td>1.00</td><td>0.97</td><td>0.95</td><td>0.94</td><td>0.94</td><td>0.96</td><td>0.99</td><td>1.03</td><td>1.05</td><td>1.07</td></tr>
<tr><td>20</td><td>1.10</td><td>1.05</td><td>0.99</td><td>0.93</td><td>0.88</td><td>0.86</td><td>0.87</td><td>0.91</td><td>0.97</td><td>1.03</td><td>1.08</td><td>1.11</td></tr>
<tr><td>30</td><td>1.11</td><td>1.04</td><td>0.95</td><td>0.86</td><td>0.79</td><td>0.76</td><td>0.78</td><td>0.83</td><td>0.92</td><td>1.01</td><td>1.09</td><td>1.13</td></tr>
<tr><td>40</td><td>1.10</td><td>1.00</td><td>0.89</td><td>0.78</td><td>0.70</td><td>0.66</td><td>0.68</td><td>0.75</td><td>0.85</td><td>0.97</td><td>1.08</td><td>1.13</td></tr>
<tr><td>50</td><td>1.06</td><td>0.95</td><td>0.82</td><td>0.69</td><td>0.59</td><td>0.55</td><td>0.57</td><td>0.65</td><td>0.77</td><td>0.92</td><td>1.04</td><td>1.10</td></tr>
<tr><td>60</td><td>1.01</td><td>0.88</td><td>0.73</td><td>0.58</td><td>0.49</td><td>0.45</td><td>0.47</td><td>0.54</td><td>0.68</td><td>0.84</td><td>0.98</td><td>1.04</td></tr>
<tr><td rowspan="6">0.5</td><td>10</td><td>1.08</td><td>1.05</td><td>1.01</td><td>0.97</td><td>0.93</td><td>0.92</td><td>0.93</td><td>0.95</td><td>0.99</td><td>1.04</td><td>1.07</td><td>1.09</td></tr>
<tr><td>20</td><td>1.13</td><td>1.07</td><td>0.99</td><td>0.91</td><td>0.85</td><td>0.82</td><td>0.83</td><td>0.89</td><td>0.96</td><td>1.05</td><td>1.12</td><td>1.15</td></tr>
<tr><td>30</td><td>1.16</td><td>1.06</td><td>0.95</td><td>0.84</td><td>0.75</td><td>0.71</td><td>0.73</td><td>0.80</td><td>0.91</td><td>1.03</td><td>1.14</td><td>1.19</td></tr>
<tr><td>40</td><td>1.16</td><td>1.03</td><td>0.89</td><td>0.74</td><td>0.63</td><td>0.59</td><td>0.61</td><td>0.70</td><td>0.84</td><td>0.99</td><td>1.13</td><td>1.19</td></tr>
<tr><td>50</td><td>1.12</td><td>0.98</td><td>0.81</td><td>0.63</td><td>0.51</td><td>0.46</td><td>0.49</td><td>0.59</td><td>0.75</td><td>0.93</td><td>1.09</td><td>1.17</td></tr>
<tr><td>60</td><td>1.07</td><td>0.90</td><td>0.71</td><td>0.52</td><td>0.39</td><td>0.34</td><td>0.37</td><td>0.47</td><td>0.64</td><td>0.85</td><td>1.03</td><td>1.12</td></tr>
<tr><td rowspan="6">0.6</td><td>10</td><td>1.10</td><td>1.06</td><td>1.01</td><td>0.96</td><td>0.92</td><td>0.90</td><td>0.91</td><td>0.94</td><td>0.99</td><td>1.05</td><td>1.09</td><td>1.11</td></tr>
<tr><td>20</td><td>1.17</td><td>1.09</td><td>0.99</td><td>0.89</td><td>0.82</td><td>0.78</td><td>0.80</td><td>0.86</td><td>0.96</td><td>1.06</td><td>1.15</td><td>1.19</td></tr>
<tr><td>30</td><td>1.21</td><td>1.09</td><td>0.95</td><td>0.81</td><td>0.70</td><td>0.65</td><td>0.67</td><td>0.77</td><td>0.90</td><td>1.05</td><td>1.18</td><td>1.25</td></tr>
<tr><td>40</td><td>1.22</td><td>1.07</td><td>0.89</td><td>0.70</td><td>0.57</td><td>0.51</td><td>0.54</td><td>0.65</td><td>0.82</td><td>1.02</td><td>1.18</td><td>1.26</td></tr>
<tr><td>50</td><td>1.19</td><td>1.01</td><td>0.80</td><td>0.58</td><td>0.43</td><td>0.36</td><td>0.40</td><td>0.52</td><td>0.72</td><td>0.95</td><td>1.15</td><td>1.25</td></tr>
<tr><td>60</td><td>1.13</td><td>0.93</td><td>0.69</td><td>0.45</td><td>0.29</td><td>0.23</td><td>0.26</td><td>0.38</td><td>0.60</td><td>0.86</td><td>1.09</td><td>1.20</td></tr>
<tr><td rowspan="6">0.7</td><td>10</td><td>1.12</td><td>1.07</td><td>1.01</td><td>0.95</td><td>0.90</td><td>0.88</td><td>0.89</td><td>0.93</td><td>0.99</td><td>1.06</td><td>1.11</td><td>1.14</td></tr>
<tr><td>20</td><td>1.21</td><td>1.12</td><td>1.00</td><td>0.88</td><td>0.78</td><td>0.74</td><td>0.76</td><td>0.84</td><td>0.96</td><td>1.08</td><td>1.19</td><td>1.24</td></tr>
<tr><td>30</td><td>1.27</td><td>1.13</td><td>0.95</td><td>0.78</td><td>0.64</td><td>0.58</td><td>0.61</td><td>0.72</td><td>0.89</td><td>1.08</td><td>1.24</td><td>1.31</td></tr>
<tr><td>40</td><td>1.29</td><td>1.10</td><td>0.88</td><td>0.65</td><td>0.49</td><td>0.42</td><td>0.45</td><td>0.59</td><td>0.80</td><td>1.04</td><td>1.25</td><td>1.34</td></tr>
<tr><td>50</td><td>1.27</td><td>1.05</td><td>0.78</td><td>0.52</td><td>0.33</td><td>0.25</td><td>0.29</td><td>0.44</td><td>0.69</td><td>0.97</td><td>1.22</td><td>1.33</td></tr>
<tr><td>60</td><td>1.21</td><td>0.96</td><td>0.66</td><td>0.37</td><td>0.17</td><td>0.10</td><td>0.14</td><td>0.29</td><td>0.55</td><td>0.88</td><td>1.15</td><td>1.29</td></tr>
<tr><td></td><td></td><td>Jul</td><td>Aug</td><td>Sep</td><td>Oct</td><td>Nov</td><td>Dec</td><td>Jan</td><td>Feb</td><td>Mar</td><td>Apr</td><td>May</td><td>Jun</td></tr>
<tr><td></td><td></td><td colspan="12">Month for Southern Hemisphere</td></tr>
</table>

Table A3  Tilt factors for latitude 5 degrees

Month for Northern Hemisphere

| Clearness Index | Tilt (degrees) | Jan | Feb | Mar | Apr | May | Jun | Jul | Aug | Sep | Oct | Nov | Dec |
|---|---|---|---|---|---|---|---|---|---|---|---|---|---|
|     | 10 | 1.05 | 1.03 | 1.01 | 0.99 | 0.97 | 0.96 | 0.96 | 0.98 | 1.00 | 1.02 | 1.04 | 1.05 |
|     | 20 | 1.07 | 1.04 | 1.00 | 0.95 | 0.92 | 0.91 | 0.91 | 0.94 | 0.98 | 1.03 | 1.07 | 1.09 |
| 0.3 | 30 | 1.08 | 1.03 | 0.97 | 0.90 | 0.86 | 0.84 | 0.85 | 0.89 | 0.94 | 1.01 | 1.07 | 1.10 |
|     | 40 | 1.07 | 1.00 | 0.92 | 0.84 | 0.79 | 0.76 | 0.77 | 0.82 | 0.89 | 0.98 | 1.05 | 1.09 |
|     | 50 | 1.03 | 0.95 | 0.86 | 0.77 | 0.70 | 0.68 | 0.69 | 0.74 | 0.82 | 0.93 | 1.02 | 1.06 |
|     | 60 | 0.98 | 0.89 | 0.78 | 0.68 | 0.61 | 0.59 | 0.60 | 0.65 | 0.75 | 0.86 | 0.96 | 1.01 |
|     | 10 | 1.07 | 1.04 | 1.01 | 0.98 | 0.96 | 0.95 | 0.95 | 0.97 | 1.00 | 1.04 | 1.07 | 1.08 |
|     | 20 | 1.12 | 1.07 | 1.00 | 0.94 | 0.90 | 0.87 | 0.88 | 0.92 | 0.98 | 1.05 | 1.11 | 1.14 |
| 0.4 | 30 | 1.14 | 1.07 | 0.98 | 0.89 | 0.82 | 0.79 | 0.80 | 0.86 | 0.94 | 1.04 | 1.13 | 1.17 |
|     | 40 | 1.14 | 1.04 | 0.93 | 0.81 | 0.73 | 0.69 | 0.71 | 0.78 | 0.89 | 1.01 | 1.12 | 1.17 |
|     | 50 | 1.12 | 1.00 | 0.86 | 0.72 | 0.63 | 0.59 | 0.61 | 0.69 | 0.81 | 0.96 | 1.09 | 1.15 |
|     | 60 | 1.07 | 0.93 | 0.78 | 0.62 | 0.53 | 0.49 | 0.51 | 0.58 | 0.72 | 0.89 | 1.04 | 1.11 |
|     | 10 | 1.09 | 1.06 | 1.02 | 0.98 | 0.94 | 0.93 | 0.94 | 0.96 | 1.00 | 1.05 | 1.09 | 1.11 |
|     | 20 | 1.16 | 1.09 | 1.01 | 0.93 | 0.87 | 0.84 | 0.86 | 0.91 | 0.98 | 1.07 | 1.15 | 1.18 |
| 0.5 | 30 | 1.20 | 1.10 | 0.98 | 0.87 | 0.78 | 0.74 | 0.76 | 0.83 | 0.94 | 1.07 | 1.18 | 1.24 |
|     | 40 | 1.21 | 1.08 | 0.93 | 0.78 | 0.68 | 0.63 | 0.65 | 0.74 | 0.88 | 1.04 | 1.18 | 1.25 |
|     | 50 | 1.19 | 1.04 | 0.86 | 0.68 | 0.56 | 0.51 | 0.53 | 0.63 | 0.80 | 0.99 | 1.16 | 1.24 |
|     | 60 | 1.15 | 0.97 | 0.77 | 0.57 | 0.44 | 0.39 | 0.42 | 0.52 | 0.70 | 0.91 | 1.11 | 1.20 |
|     | 10 | 1.12 | 1.08 | 1.02 | 0.97 | 0.93 | 0.91 | 0.92 | 0.96 | 1.01 | 1.06 | 1.11 | 1.13 |
|     | 20 | 1.21 | 1.12 | 1.02 | 0.92 | 0.84 | 0.81 | 0.83 | 0.89 | 0.99 | 1.09 | 1.19 | 1.24 |
| 0.6 | 30 | 1.27 | 1.14 | 0.99 | 0.85 | 0.74 | 0.69 | 0.71 | 0.81 | 0.94 | 1.10 | 1.24 | 1.31 |
|     | 40 | 1.29 | 1.13 | 0.94 | 0.75 | 0.62 | 0.56 | 0.59 | 0.70 | 0.87 | 1.07 | 1.25 | 1.34 |
|     | 50 | 1.28 | 1.08 | 0.86 | 0.64 | 0.49 | 0.42 | 0.46 | 0.58 | 0.78 | 1.02 | 1.23 | 1.34 |
|     | 60 | 1.23 | 1.01 | 0.76 | 0.51 | 0.35 | 0.29 | 0.32 | 0.45 | 0.67 | 0.94 | 1.18 | 1.30 |
|     | 10 | 1.15 | 1.09 | 1.03 | 0.97 | 0.92 | 0.90 | 0.91 | 0.95 | 1.01 | 1.08 | 1.14 | 1.17 |
|     | 20 | 1.26 | 1.16 | 1.03 | 0.91 | 0.81 | 0.77 | 0.79 | 0.87 | 0.99 | 1.12 | 1.24 | 1.30 |
| 0.7 | 30 | 1.34 | 1.18 | 1.01 | 0.82 | 0.69 | 0.63 | 0.66 | 0.77 | 0.94 | 1.13 | 1.31 | 1.39 |
|     | 40 | 1.38 | 1.18 | 0.95 | 0.72 | 0.55 | 0.48 | 0.52 | 0.65 | 0.86 | 1.11 | 1.33 | 1.44 |
|     | 50 | 1.37 | 1.14 | 0.86 | 0.59 | 0.40 | 0.33 | 0.36 | 0.52 | 0.76 | 1.06 | 1.32 | 1.45 |
|     | 60 | 1.33 | 1.06 | 0.75 | 0.45 | 0.25 | 0.17 | 0.21 | 0.37 | 0.64 | 0.97 | 1.27 | 1.42 |
|     |     | Jul | Aug | Sep | Oct | Nov | Dec | Jan | Feb | Mar | Apr | May | Jun |

Month for Southern Hemisphere

Table A4  Tilt factors for latitude 10 degrees

| Clearness Index | Tilt (degrees) | Jan | Feb | Mar | Apr | May | Jun | Jul | Aug | Sep | Oct | Nov | Dec |
|---|---|---|---|---|---|---|---|---|---|---|---|---|---|
| | | | | | | Month for Northern Hemisphere | | | | | | | |
| 0.3 | 10 | 1.06 | 1.04 | 1.01 | 0.99 | 0.97 | 0.97 | 0.97 | 0.98 | 1.01 | 1.03 | 1.05 | 1.06 |
| | 20 | 1.09 | 1.05 | 1.01 | 0.96 | 0.93 | 0.92 | 0.92 | 0.95 | 0.99 | 1.04 | 1.08 | 1.11 |
| | 30 | 1.11 | 1.05 | 0.98 | 0.92 | 0.88 | 0.86 | 0.87 | 0.90 | 0.96 | 1.03 | 1.10 | 1.13 |
| | 40 | 1.10 | 1.03 | 0.94 | 0.86 | 0.81 | 0.78 | 0.79 | 0.84 | 0.91 | 1.00 | 1.09 | 1.13 |
| | 50 | 1.08 | 0.99 | 0.89 | 0.79 | 0.73 | 0.70 | 0.71 | 0.77 | 0.85 | 0.96 | 1.06 | 1.10 |
| | 60 | 1.03 | 0.93 | 0.81 | 0.71 | 0.64 | 0.61 | 0.62 | 0.68 | 0.77 | 0.90 | 1.01 | 1.06 |
| 0.4 | 10 | 1.09 | 1.06 | 1.02 | 0.99 | 0.96 | 0.95 | 0.96 | 0.98 | 1.01 | 1.05 | 1.08 | 1.09 |
| | 20 | 1.15 | 1.09 | 1.02 | 0.96 | 0.91 | 0.89 | 0.90 | 0.94 | 1.00 | 1.07 | 1.13 | 1.17 |
| | 30 | 1.18 | 1.10 | 1.00 | 0.91 | 0.84 | 0.81 | 0.83 | 0.88 | 0.97 | 1.07 | 1.16 | 1.21 |
| | 40 | 1.19 | 1.08 | 0.96 | 0.84 | 0.76 | 0.72 | 0.74 | 0.81 | 0.92 | 1.05 | 1.17 | 1.23 |
| | 50 | 1.18 | 1.05 | 0.90 | 0.76 | 0.67 | 0.63 | 0.65 | 0.72 | 0.85 | 1.00 | 1.15 | 1.22 |
| | 60 | 1.14 | 0.99 | 0.82 | 0.66 | 0.56 | 0.52 | 0.54 | 0.62 | 0.76 | 0.94 | 1.10 | 1.18 |
| 0.5 | 10 | 1.11 | 1.07 | 1.03 | 0.99 | 0.96 | 0.94 | 0.95 | 0.98 | 1.02 | 1.06 | 1.10 | 1.13 |
| | 20 | 1.2 | 1.12 | 1.04 | 0.95 | 0.89 | 0.87 | 0.88 | 0.93 | 1.01 | 1.10 | 1.18 | 1.22 |
| | 30 | 1.26 | 1.14 | 1.02 | 0.90 | 0.81 | 0.77 | 0.79 | 0.86 | 0.98 | 1.11 | 1.23 | 1.29 |
| | 40 | 1.28 | 1.14 | 0.98 | 0.82 | 0.71 | 0.67 | 0.69 | 0.78 | 0.92 | 1.09 | 1.25 | 1.33 |
| | 50 | 1.28 | 1.10 | 0.91 | 0.73 | 0.61 | 0.56 | 0.58 | 0.68 | 0.85 | 1.05 | 1.24 | 1.33 |
| | 60 | 1.24 | 1.04 | 0.83 | 0.62 | 0.49 | 0.44 | 0.47 | 0.57 | 0.75 | 0.98 | 1.19 | 1.30 |
| 0.6 | 10 | 1.14 | 1.09 | 1.04 | 0.99 | 0.95 | 0.93 | 0.94 | 0.97 | 1.02 | 1.08 | 1.13 | 1.16 |
| | 20 | 1.25 | 1.16 | 1.05 | 0.95 | 0.87 | 0.84 | 0.85 | 0.92 | 1.01 | 1.13 | 1.23 | 1.28 |
| | 30 | 1.33 | 1.19 | 1.04 | 0.89 | 0.78 | 0.73 | 0.75 | 0.84 | 0.98 | 1.15 | 1.30 | 1.38 |
| | 40 | 1.37 | 1.20 | 1.00 | 0.80 | 0.67 | 0.61 | 0.64 | 0.75 | 0.92 | 1.14 | 1.33 | 1.43 |
| | 50 | 1.38 | 1.17 | 0.93 | 0.70 | 0.55 | 0.48 | 0.51 | 0.64 | 0.84 | 1.10 | 1.33 | 1.45 |
| | 60 | 1.35 | 1.10 | 0.84 | 0.58 | 0.42 | 0.35 | 0.38 | 0.51 | 0.74 | 1.03 | 1.29 | 1.42 |
| 0.7 | 10 | 1.18 | 1.12 | 1.05 | 0.98 | 0.94 | 0.91 | 0.92 | 0.97 | 1.03 | 1.10 | 1.16 | 1.20 |
| | 20 | 1.32 | 1.20 | 1.07 | 0.94 | 0.85 | 0.81 | 0.83 | 0.91 | 1.02 | 1.16 | 1.29 | 1.35 |
| | 30 | 1.42 | 1.25 | 1.06 | 0.87 | 0.74 | 0.68 | 0.71 | 0.82 | 0.99 | 1.19 | 1.38 | 1.47 |
| | 40 | 1.48 | 1.26 | 1.02 | 0.78 | 0.61 | 0.54 | 0.58 | 0.71 | 0.93 | 1.19 | 1.43 | 1.55 |
| | 50 | 1.50 | 1.24 | 0.94 | 0.66 | 0.47 | 0.40 | 0.43 | 0.59 | 0.84 | 1.15 | 1.44 | 1.58 |
| | 60 | 1.47 | 1.17 | 0.84 | 0.53 | 0.33 | 0.25 | 0.29 | 0.45 | 0.73 | 1.08 | 1.40 | 1.56 |
| | | Jul | Aug | Sep | Oct | Nov | Dec | Jan | Feb | Mar | Apr | May | Jun |
| | | | | | Month for Southern Hemisphere | | | | | | | | |

Table A5  Tilt factor for latitude 15 degrees

<table>
<tr><td rowspan="2">Clearness Index</td><td rowspan="2">Tilt (degrees)</td><td colspan="12" align="center">Month for Northern Hemisphere</td></tr>
<tr><td>Jan</td><td>Feb</td><td>Mar</td><td>Apr</td><td>May</td><td>Jun</td><td>Jul</td><td>Aug</td><td>Sep</td><td>Oct</td><td>Nov</td><td>Dec</td></tr>
<tr><td rowspan="6">0.3</td><td>10</td><td>1.07</td><td>1.04</td><td>1.02</td><td>1.00</td><td>0.98</td><td>0.97</td><td>0.98</td><td>0.99</td><td>1.01</td><td>1.04</td><td>1.06</td><td>1.07</td></tr>
<tr><td>20</td><td>1.11</td><td>1.07</td><td>1.02</td><td>0.98</td><td>0.94</td><td>0.93</td><td>0.94</td><td>0.96</td><td>1.00</td><td>1.06</td><td>1.10</td><td>1.13</td></tr>
<tr><td>30</td><td>1.14</td><td>1.07</td><td>1.00</td><td>0.94</td><td>0.89</td><td>0.87</td><td>0.88</td><td>0.92</td><td>0.98</td><td>1.05</td><td>1.13</td><td>1.16</td></tr>
<tr><td>40</td><td>1.14</td><td>1.06</td><td>0.97</td><td>0.88</td><td>0.83</td><td>0.80</td><td>0.81</td><td>0.86</td><td>0.94</td><td>1.03</td><td>1.12</td><td>1.17</td></tr>
<tr><td>50</td><td>1.13</td><td>1.02</td><td>0.92</td><td>0.82</td><td>0.75</td><td>0.72</td><td>0.74</td><td>0.79</td><td>0.88</td><td>0.99</td><td>1.10</td><td>1.16</td></tr>
<tr><td>60</td><td>1.08</td><td>0.97</td><td>0.85</td><td>0.74</td><td>0.66</td><td>0.64</td><td>0.65</td><td>0.71</td><td>0.81</td><td>0.94</td><td>1.06</td><td>1.12</td></tr>
<tr><td rowspan="6">0.4</td><td>10</td><td>1.10</td><td>1.07</td><td>1.03</td><td>1.00</td><td>0.97</td><td>0.96</td><td>0.97</td><td>0.99</td><td>1.02</td><td>1.06</td><td>1.09</td><td>1.11</td></tr>
<tr><td>20</td><td>1.18</td><td>1.11</td><td>1.04</td><td>0.98</td><td>0.93</td><td>0.91</td><td>0.92</td><td>0.96</td><td>1.02</td><td>1.09</td><td>1.16</td><td>1.20</td></tr>
<tr><td>30</td><td>1.23</td><td>1.13</td><td>1.03</td><td>0.93</td><td>0.87</td><td>0.84</td><td>0.85</td><td>0.91</td><td>0.99</td><td>1.10</td><td>1.21</td><td>1.26</td></tr>
<tr><td>40</td><td>1.25</td><td>1.13</td><td>1.00</td><td>0.87</td><td>0.79</td><td>0.75</td><td>0.77</td><td>0.84</td><td>0.95</td><td>1.09</td><td>1.23</td><td>1.29</td></tr>
<tr><td>50</td><td>1.25</td><td>1.10</td><td>0.94</td><td>0.80</td><td>0.70</td><td>0.66</td><td>0.68</td><td>0.76</td><td>0.89</td><td>1.06</td><td>1.22</td><td>1.30</td></tr>
<tr><td>60</td><td>1.22</td><td>1.05</td><td>0.87</td><td>0.71</td><td>0.60</td><td>0.56</td><td>0.58</td><td>0.66</td><td>0.81</td><td>1.00</td><td>1.18</td><td>1.27</td></tr>
<tr><td rowspan="6">0.5</td><td>10</td><td>1.13</td><td>1.09</td><td>1.04</td><td>1.00</td><td>0.97</td><td>0.95</td><td>0.96</td><td>0.99</td><td>1.03</td><td>1.08</td><td>1.12</td><td>1.15</td></tr>
<tr><td>20</td><td>1.24</td><td>1.16</td><td>1.06</td><td>0.97</td><td>0.91</td><td>0.89</td><td>0.90</td><td>0.95</td><td>1.03</td><td>1.13</td><td>1.22</td><td>1.27</td></tr>
<tr><td>30</td><td>1.32</td><td>1.19</td><td>1.06</td><td>0.93</td><td>0.84</td><td>0.80</td><td>0.82</td><td>0.89</td><td>1.01</td><td>1.15</td><td>1.29</td><td>1.36</td></tr>
<tr><td>40</td><td>1.36</td><td>1.20</td><td>1.02</td><td>0.86</td><td>0.75</td><td>0.71</td><td>0.73</td><td>0.82</td><td>0.97</td><td>1.15</td><td>1.32</td><td>1.41</td></tr>
<tr><td>50</td><td>1.37</td><td>1.18</td><td>0.97</td><td>0.78</td><td>0.65</td><td>0.60</td><td>0.63</td><td>0.73</td><td>0.90</td><td>1.12</td><td>1.32</td><td>1.43</td></tr>
<tr><td>60</td><td>1.34</td><td>1.13</td><td>0.89</td><td>0.68</td><td>0.54</td><td>0.49</td><td>0.51</td><td>0.62</td><td>0.81</td><td>1.06</td><td>1.29</td><td>1.41</td></tr>
<tr><td rowspan="6">0.6</td><td>10</td><td>1.17</td><td>1.11</td><td>1.06</td><td>1.00</td><td>0.96</td><td>0.94</td><td>0.95</td><td>0.98</td><td>1.04</td><td>1.10</td><td>1.16</td><td>1.19</td></tr>
<tr><td>20</td><td>1.31</td><td>1.2</td><td>1.08</td><td>0.97</td><td>0.90</td><td>0.86</td><td>0.88</td><td>0.94</td><td>1.04</td><td>1.17</td><td>1.28</td><td>1.34</td></tr>
<tr><td>30</td><td>1.41</td><td>1.25</td><td>1.08</td><td>0.92</td><td>0.81</td><td>0.77</td><td>0.79</td><td>0.88</td><td>1.03</td><td>1.20</td><td>1.37</td><td>1.46</td></tr>
<tr><td>40</td><td>1.47</td><td>1.27</td><td>1.05</td><td>0.85</td><td>0.72</td><td>0.66</td><td>0.69</td><td>0.80</td><td>0.98</td><td>1.21</td><td>1.43</td><td>1.54</td></tr>
<tr><td>50</td><td>1.49</td><td>1.26</td><td>1.00</td><td>0.76</td><td>0.60</td><td>0.54</td><td>0.57</td><td>0.70</td><td>0.91</td><td>1.18</td><td>1.44</td><td>1.57</td></tr>
<tr><td>60</td><td>1.48</td><td>1.21</td><td>0.91</td><td>0.65</td><td>0.48</td><td>0.41</td><td>0.44</td><td>0.58</td><td>0.81</td><td>1.12</td><td>1.41</td><td>1.56</td></tr>
<tr><td rowspan="6">0.7</td><td>10</td><td>1.21</td><td>1.14</td><td>1.07</td><td>1.00</td><td>0.95</td><td>0.93</td><td>0.94</td><td>0.98</td><td>1.05</td><td>1.12</td><td>1.19</td><td>1.23</td></tr>
<tr><td>20</td><td>1.38</td><td>1.25</td><td>1.11</td><td>0.97</td><td>0.88</td><td>0.84</td><td>0.86</td><td>0.94</td><td>1.06</td><td>1.21</td><td>1.35</td><td>1.42</td></tr>
<tr><td>30</td><td>1.51</td><td>1.32</td><td>1.11</td><td>0.92</td><td>0.78</td><td>0.73</td><td>0.76</td><td>0.87</td><td>1.04</td><td>1.26</td><td>1.47</td><td>1.58</td></tr>
<tr><td>40</td><td>1.60</td><td>1.35</td><td>1.09</td><td>0.84</td><td>0.67</td><td>0.60</td><td>0.64</td><td>0.77</td><td>1.00</td><td>1.28</td><td>1.54</td><td>1.68</td></tr>
<tr><td>50</td><td>1.64</td><td>1.35</td><td>1.03</td><td>0.74</td><td>0.54</td><td>0.46</td><td>0.50</td><td>0.66</td><td>0.92</td><td>1.25</td><td>1.57</td><td>1.74</td></tr>
<tr><td>60</td><td>1.63</td><td>1.30</td><td>0.94</td><td>0.61</td><td>0.41</td><td>0.32</td><td>0.36</td><td>0.53</td><td>0.82</td><td>1.20</td><td>1.56</td><td>1.74</td></tr>
<tr><td></td><td></td><td>Jul</td><td>Aug</td><td>Sep</td><td>Oct</td><td>Nov</td><td>Dec</td><td>Jan</td><td>Feb</td><td>Mar</td><td>Apr</td><td>May</td><td>Jun</td></tr>
<tr><td></td><td></td><td colspan="12" align="center">Month for Southern Hemisphere</td></tr>
</table>

Table A6 Tilt factors for latitude 20 degrees

Month for Northern Hemisphere

| Clearness Index | Tilt (degrees) | Jan | Feb | Mar | Apr | May | Jun | Jul | Aug | Sep | Oct | Nov | Dec |
|---|---|---|---|---|---|---|---|---|---|---|---|---|---|
| 0.3 | 10 | 1.08 | 1.05 | 1.03 | 1.00 | 0.98 | 0.98 | 0.98 | 1.00 | 1.02 | 1.05 | 1.07 | 1.09 |
| | 20 | 1.14 | 1.09 | 1.04 | 0.99 | 0.95 | 0.94 | 0.95 | 0.97 | 1.02 | 1.07 | 1.13 | 1.16 |
| | 30 | 1.18 | 1.10 | 1.02 | 0.95 | 0.91 | 0.89 | 0.90 | 0.93 | 1.00 | 1.08 | 1.16 | 1.20 |
| | 40 | 1.19 | 1.10 | 1.00 | 0.90 | 0.85 | 0.82 | 0.83 | 0.88 | 0.96 | 1.07 | 1.17 | 1.23 |
| | 50 | 1.18 | 1.07 | 0.95 | 0.84 | 0.77 | 0.75 | 0.76 | 0.81 | 0.91 | 1.03 | 1.16 | 1.22 |
| | 60 | 1.15 | 1.02 | 0.88 | 0.76 | 0.69 | 0.66 | 0.68 | 0.73 | 0.84 | 0.98 | 1.12 | 1.19 |
| 0.4 | 10 | 1.12 | 1.08 | 1.04 | 1.01 | 0.98 | 0.97 | 0.97 | 1.00 | 1.03 | 1.07 | 1.11 | 1.13 |
| | 20 | 1.22 | 1.14 | 1.06 | 0.99 | 0.94 | 0.92 | 0.93 | 0.97 | 1.04 | 1.12 | 1.20 | 1.24 |
| | 30 | 1.29 | 1.18 | 1.06 | 0.96 | 0.89 | 0.86 | 0.87 | 0.93 | 1.02 | 1.14 | 1.26 | 1.32 |
| | 40 | 1.33 | 1.18 | 1.04 | 0.90 | 0.82 | 0.78 | 0.80 | 0.87 | 0.99 | 1.14 | 1.29 | 1.37 |
| | 50 | 1.34 | 1.17 | 0.99 | 0.83 | 0.73 | 0.69 | 0.71 | 0.79 | 0.93 | 1.11 | 1.30 | 1.39 |
| | 60 | 1.31 | 1.12 | 0.92 | 0.75 | 0.64 | 0.60 | 0.62 | 0.70 | 0.86 | 1.06 | 1.27 | 1.38 |
| 0.5 | 10 | 1.16 | 1.11 | 1.06 | 1.01 | 0.98 | 0.96 | 0.97 | 1.00 | 1.04 | 1.09 | 1.15 | 1.18 |
| | 20 | 1.29 | 1.19 | 1.09 | 1.00 | 0.93 | 0.91 | 0.92 | 0.97 | 1.06 | 1.16 | 1.27 | 1.32 |
| | 30 | 1.39 | 1.25 | 1.10 | 0.96 | 0.87 | 0.83 | 0.85 | 0.92 | 1.05 | 1.20 | 1.36 | 1.44 |
| | 40 | 1.45 | 1.27 | 1.08 | 0.90 | 0.79 | 0.74 | 0.77 | 0.86 | 1.01 | 1.21 | 1.41 | 1.52 |
| | 50 | 1.48 | 1.26 | 1.03 | 0.83 | 0.70 | 0.64 | 0.67 | 0.77 | 0.95 | 1.19 | 1.43 | 1.55 |
| | 60 | 1.47 | 1.22 | 0.96 | 0.73 | 0.59 | 0.54 | 0.56 | 0.68 | 0.87 | 1.14 | 1.41 | 1.55 |
| 0.6 | 10 | 1.20 | 1.14 | 1.07 | 1.01 | 0.97 | 0.95 | 0.96 | 1.00 | 1.05 | 1.12 | 1.19 | 1.22 |
| | 20 | 1.37 | 1.25 | 1.12 | 1.00 | 0.92 | 0.89 | 0.90 | 0.97 | 1.07 | 1.21 | 1.34 | 1.41 |
| | 30 | 1.50 | 1.32 | 1.13 | 0.96 | 0.85 | 0.80 | 0.83 | 0.92 | 1.07 | 1.26 | 1.46 | 1.56 |
| | 40 | 1.59 | 1.36 | 1.12 | 0.90 | 0.76 | 0.70 | 0.73 | 0.85 | 1.04 | 1.29 | 1.53 | 1.67 |
| | 50 | 1.63 | 1.36 | 1.07 | 0.82 | 0.66 | 0.59 | 0.62 | 0.75 | 0.98 | 1.27 | 1.57 | 1.73 |
| | 60 | 1.63 | 1.32 | 1.00 | 0.72 | 0.54 | 0.47 | 0.50 | 0.64 | 0.89 | 1.23 | 1.56 | 1.74 |
| 0.7 | 10 | 1.25 | 1.17 | 1.09 | 1.02 | 0.97 | 0.94 | 0.96 | 1.00 | 1.06 | 1.15 | 1.23 | 1.27 |
| | 20 | 1.46 | 1.31 | 1.15 | 1.01 | 0.91 | 0.87 | 0.89 | 0.97 | 1.10 | 1.26 | 1.42 | 1.51 |
| | 30 | 1.63 | 1.40 | 1.18 | 0.97 | 0.83 | 0.77 | 0.80 | 0.91 | 1.10 | 1.34 | 1.57 | 1.70 |
| | 40 | 1.74 | 1.46 | 1.17 | 0.90 | 0.73 | 0.66 | 0.69 | 0.83 | 1.07 | 1.37 | 1.68 | 1.84 |
| | 50 | 1.81 | 1.47 | 1.12 | 0.81 | 0.61 | 0.53 | 0.57 | 0.73 | 1.00 | 1.37 | 1.73 | 1.93 |
| | 60 | 1.83 | 1.44 | 1.05 | 0.70 | 0.48 | 0.40 | 0.44 | 0.61 | 0.91 | 1.33 | 1.74 | 1.95 |
| | | Jul | Aug | Sep | Oct | Nov | Dec | Jan | Feb | Mar | Apr | May | Jun |

Month for Southern Hemisphere

Table A7  Tilt factors for latitude 25 degrees

| Clearness Index | Tilt (degrees) | Jan | Feb | Mar | Apr | May | Jun | Jul | Aug | Sep | Oct | Nov | Dec |
|---|---|---|---|---|---|---|---|---|---|---|---|---|---|
| | | | | | | Month for Northern Hemisphere | | | | | | | |
| 0.3 | 10 | 1.10 | 1.07 | 1.04 | 1.01 | 0.99 | 0.98 | 0.99 | 1.00 | 1.03 | 1.06 | 1.09 | 1.11 |
| | 20 | 1.17 | 1.11 | 1.05 | 1.00 | 0.96 | 0.95 | 0.96 | 0.98 | 1.03 | 1.09 | 1.16 | 1.19 |
| | 30 | 1.23 | 1.14 | 1.05 | 0.97 | 0.92 | 0.90 | 0.91 | 0.95 | 1.02 | 1.11 | 1.20 | 1.26 |
| | 40 | 1.25 | 1.14 | 1.02 | 0.93 | 0.86 | 0.84 | 0.85 | 0.90 | 0.99 | 1.10 | 1.23 | 1.29 |
| | 50 | 1.26 | 1.12 | 0.98 | 0.87 | 0.80 | 0.77 | 0.78 | 0.84 | 0.94 | 1.08 | 1.22 | 1.30 |
| | 60 | 1.23 | 1.08 | 0.92 | 0.79 | 0.72 | 0.69 | 0.70 | 0.76 | 0.87 | 1.03 | 1.20 | 1.29 |
| 0.4 | 10 | 1.14 | 1.10 | 1.05 | 1.01 | 0.99 | 0.98 | 0.98 | 1.00 | 1.04 | 1.09 | 1.13 | 1.16 |
| | 20 | 1.26 | 1.17 | 1.09 | 1.01 | 0.96 | 0.94 | 0.95 | 0.99 | 1.06 | 1.15 | 1.24 | 1.30 |
| | 30 | 1.36 | 1.22 | 1.09 | 0.98 | 0.91 | 0.88 | 0.89 | 0.95 | 1.05 | 1.19 | 1.32 | 1.40 |
| | 40 | 1.41 | 1.25 | 1.08 | 0.94 | 0.85 | 0.81 | 0.83 | 0.90 | 1.03 | 1.20 | 1.37 | 1.47 |
| | 50 | 1.44 | 1.24 | 1.04 | 0.87 | 0.77 | 0.73 | 0.75 | 0.83 | 0.98 | 1.18 | 1.39 | 1.51 |
| | 60 | 1.43 | 1.21 | 0.98 | 0.79 | 0.68 | 0.63 | 0.65 | 0.74 | 0.91 | 1.14 | 1.38 | 1.51 |
| 0.5 | 10 | 1.19 | 1.13 | 1.07 | 1.02 | 0.99 | 0.97 | 0.98 | 1.01 | 1.05 | 1.11 | 1.18 | 1.21 |
| | 20 | 1.35 | 1.23 | 1.12 | 1.02 | 0.95 | 0.92 | 0.94 | 0.99 | 1.08 | 1.20 | 1.32 | 1.39 |
| | 30 | 1.48 | 1.31 | 1.14 | 0.99 | 0.90 | 0.86 | 0.88 | 0.96 | 1.08 | 1.26 | 1.44 | 1.54 |
| | 40 | 1.57 | 1.35 | 1.13 | 0.95 | 0.83 | 0.78 | 0.80 | 0.90 | 1.06 | 1.28 | 1.52 | 1.64 |
| | 50 | 1.62 | 1.36 | 1.10 | 0.88 | 0.74 | 0.69 | 0.71 | 0.82 | 1.01 | 1.28 | 1.56 | 1.71 |
| | 60 | 1.62 | 1.33 | 1.04 | 0.79 | 0.64 | 0.58 | 0.61 | 0.73 | 0.94 | 1.24 | 1.55 | 1.73 |
| 0.6 | 10 | 1.24 | 1.16 | 1.09 | 1.03 | 0.98 | 0.97 | 0.97 | 1.01 | 1.07 | 1.14 | 1.22 | 1.26 |
| | 20 | 1.44 | 1.30 | 1.15 | 1.03 | 0.95 | 0.91 | 0.93 | 1.00 | 1.11 | 1.26 | 1.41 | 1.49 |
| | 30 | 1.61 | 1.40 | 1.19 | 1.00 | 0.89 | 0.84 | 0.86 | 0.96 | 1.12 | 1.33 | 1.56 | 1.68 |
| | 40 | 1.73 | 1.46 | 1.19 | 0.95 | 0.81 | 0.75 | 0.78 | 0.90 | 1.10 | 1.38 | 1.67 | 1.83 |
| | 50 | 1.80 | 1.48 | 1.16 | 0.88 | 0.71 | 0.64 | 0.68 | 0.81 | 1.05 | 1.38 | 1.73 | 1.92 |
| | 60 | 1.83 | 1.46 | 1.09 | 0.79 | 0.60 | 0.53 | 0.56 | 0.71 | 0.98 | 1.35 | 1.74 | 1.95 |
| 0.7 | 10 | 1.30 | 1.20 | 1.11 | 1.04 | 0.98 | 0.96 | 0.97 | 1.01 | 1.08 | 1.18 | 1.27 | 1.33 |
| | 20 | 1.55 | 1.37 | 1.20 | 1.04 | 0.94 | 0.90 | 0.92 | 1.00 | 1.14 | 1.32 | 1.51 | 1.61 |
| | 30 | 1.76 | 1.50 | 1.24 | 1.02 | 0.87 | 0.81 | 0.84 | 0.96 | 1.16 | 1.42 | 1.70 | 1.85 |
| | 40 | 1.92 | 1.58 | 1.25 | 0.97 | 0.78 | 0.71 | 0.75 | 0.89 | 1.14 | 1.48 | 1.84 | 2.04 |
| | 50 | 2.02 | 1.62 | 1.22 | 0.89 | 0.68 | 0.59 | 0.63 | 0.80 | 1.10 | 1.50 | 1.93 | 2.16 |
| | 60 | 2.06 | 1.61 | 1.16 | 0.78 | 0.56 | 0.47 | 0.51 | 0.69 | 1.02 | 1.48 | 1.95 | 2.22 |
| | | Jul | Aug | Sep | Oct | Nov | Dec | Jan | Feb | Mar | Apr | May | Jun |
| | | | | | | Month for Southern Hemisphere | | | | | | | |

Table A8  Tilt factors for latitude 35 degrees

| Clearness Index | Tilt (degrees) | Jan | Feb | Mar | Apr | May | Jun | Jul | Aug | Sep | Oct | Nov | Dec |
|---|---|---|---|---|---|---|---|---|---|---|---|---|---|
| 0.3 | 10 | 1.12 | 1.08 | 1.04 | 1.01 | 0.99 | 0.99 | 0.99 | 1.01 | 1.03 | 1.07 | 1.11 | 1.13 |
| | 20 | 1.21 | 1.14 | 1.07 | 1.01 | 0.97 | 0.96 | 0.97 | 1.00 | 1.05 | 1.12 | 1.20 | 1.24 |
| | 30 | 1.29 | 1.18 | 1.07 | 0.99 | 0.94 | 0.91 | 0.93 | 0.97 | 1.04 | 1.14 | 1.26 | 1.32 |
| | 40 | 1.33 | 1.19 | 1.06 | 0.95 | 0.88 | 0.86 | 0.87 | 0.92 | 1.02 | 1.15 | 1.30 | 1.38 |
| | 50 | 1.35 | 1.18 | 1.02 | 0.89 | 0.82 | 0.79 | 0.80 | 0.86 | 0.97 | 1.13 | 1.31 | 1.41 |
| | 60 | 1.34 | 1.15 | 0.97 | 0.82 | 0.74 | 0.71 | 0.73 | 0.79 | 0.91 | 1.09 | 1.29 | 1.40 |
| 0.4 | 10 | 1.18 | 1.12 | 1.07 | 1.02 | 1.00 | 0.98 | 0.99 | 1.01 | 1.05 | 1.10 | 1.16 | 1.20 |
| | 20 | 1.32 | 1.21 | 1.11 | 1.03 | 0.97 | 0.95 | 0.96 | 1.00 | 1.08 | 1.18 | 1.30 | 1.36 |
| | 30 | 1.44 | 1.28 | 1.13 | 1.01 | 0.93 | 0.90 | 0.92 | 0.98 | 1.08 | 1.24 | 1.40 | 1.50 |
| | 40 | 1.53 | 1.32 | 1.13 | 0.97 | 0.87 | 0.83 | 0.85 | 0.93 | 1.07 | 1.26 | 1.48 | 1.60 |
| | 50 | 1.57 | 1.33 | 1.10 | 0.91 | 0.80 | 0.76 | 0.78 | 0.87 | 1.03 | 1.26 | 1.52 | 1.66 |
| | 60 | 1.58 | 1.31 | 1.05 | 0.84 | 0.71 | 0.67 | 0.69 | 0.79 | 0.96 | 1.23 | 1.52 | 1.68 |
| 0.5 | 10 | 1.23 | 1.16 | 1.09 | 1.03 | 1.00 | 0.98 | 0.99 | 1.02 | 1.07 | 1.14 | 1.21 | 1.26 |
| | 20 | 1.43 | 1.29 | 1.15 | 1.04 | 0.97 | 0.94 | 0.96 | 1.01 | 1.11 | 1.25 | 1.39 | 1.48 |
| | 30 | 1.59 | 1.38 | 1.19 | 1.03 | 0.93 | 0.89 | 0.91 | 0.99 | 1.13 | 1.32 | 1.54 | 1.67 |
| | 40 | 1.72 | 1.45 | 1.19 | 0.99 | 0.86 | 0.81 | 0.84 | 0.94 | 1.12 | 1.37 | 1.65 | 1.81 |
| | 50 | 1.79 | 1.47 | 1.17 | 0.93 | 0.78 | 0.73 | 0.75 | 0.87 | 1.08 | 1.38 | 1.71 | 1.91 |
| | 60 | 1.82 | 1.46 | 1.12 | 0.85 | 0.69 | 0.63 | 0.66 | 0.78 | 1.01 | 1.36 | 1.73 | 1.95 |
| 0.6 | 10 | 1.29 | 1.2 | 1.11 | 1.04 | 1.00 | 0.98 | 0.99 | 1.02 | 1.09 | 1.17 | 1.27 | 1.32 |
| | 20 | 1.54 | 1.36 | 1.20 | 1.06 | 0.97 | 0.93 | 0.95 | 1.02 | 1.14 | 1.31 | 1.50 | 1.61 |
| | 30 | 1.75 | 1.49 | 1.25 | 1.05 | 0.92 | 0.87 | 0.90 | 1.00 | 1.17 | 1.42 | 1.69 | 1.85 |
| | 40 | 1.91 | 1.58 | 1.27 | 1.01 | 0.85 | 0.79 | 0.82 | 0.95 | 1.17 | 1.48 | 1.83 | 2.03 |
| | 50 | 2.02 | 1.62 | 1.25 | 0.95 | 0.77 | 0.69 | 0.73 | 0.87 | 1.13 | 1.51 | 1.92 | 2.17 |
| | 60 | 2.07 | 1.62 | 1.20 | 0.86 | 0.66 | 0.59 | 0.62 | 0.78 | 1.07 | 1.49 | 1.96 | 2.24 |
| 0.7 | 10 | 1.30 | 1.20 | 1.11 | 1.04 | 0.98 | 0.96 | 0.97 | 1.01 | 1.08 | 1.18 | 1.27 | 1.33 |
| | 20 | 1.55 | 1.37 | 1.20 | 1.04 | 0.94 | 0.90 | 0.92 | 1.00 | 1.14 | 1.32 | 1.51 | 1.61 |
| | 30 | 1.76 | 1.50 | 1.24 | 1.02 | 0.87 | 0.81 | 0.84 | 0.96 | 1.16 | 1.42 | 1.70 | 1.85 |
| | 40 | 1.92 | 1.58 | 1.25 | 0.97 | 0.78 | 0.71 | 0.75 | 0.89 | 1.14 | 1.48 | 1.84 | 2.04 |
| | 50 | 2.02 | 1.62 | 1.22 | 0.89 | 0.68 | 0.59 | 0.63 | 0.80 | 1.10 | 1.50 | 1.93 | 2.16 |
| | 60 | 2.06 | 1.61 | 1.16 | 0.78 | 0.56 | 0.47 | 0.51 | 0.69 | 1.02 | 1.48 | 1.95 | 2.22 |

Month for Northern Hemisphere

| Jul | Aug | Sep | Oct | Nov | Dec | Jan | Feb | Mar | Apr | May | Jun |
|---|---|---|---|---|---|---|---|---|---|---|---|

Month for Southern Hemisphere

Table A9   Tilt factors for latitude 40 degrees

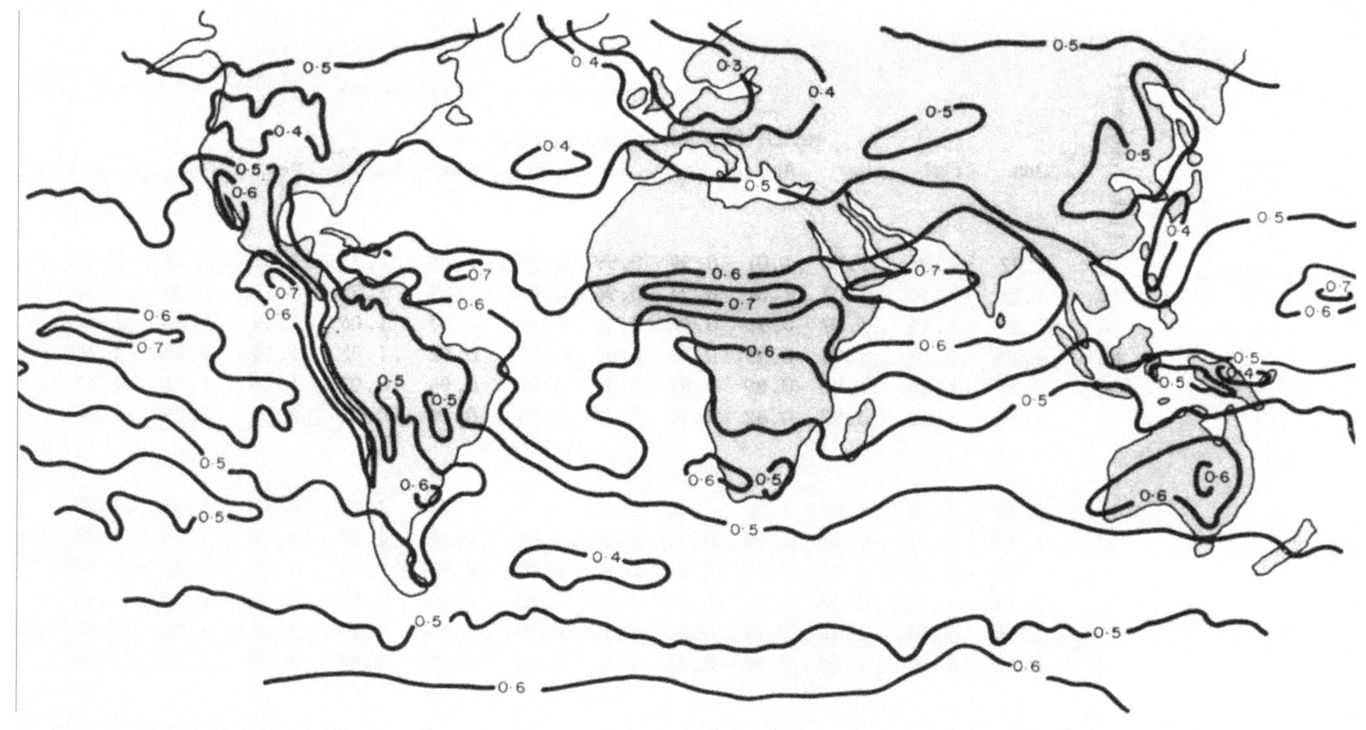

Figure A1. Clearness Index Map for January

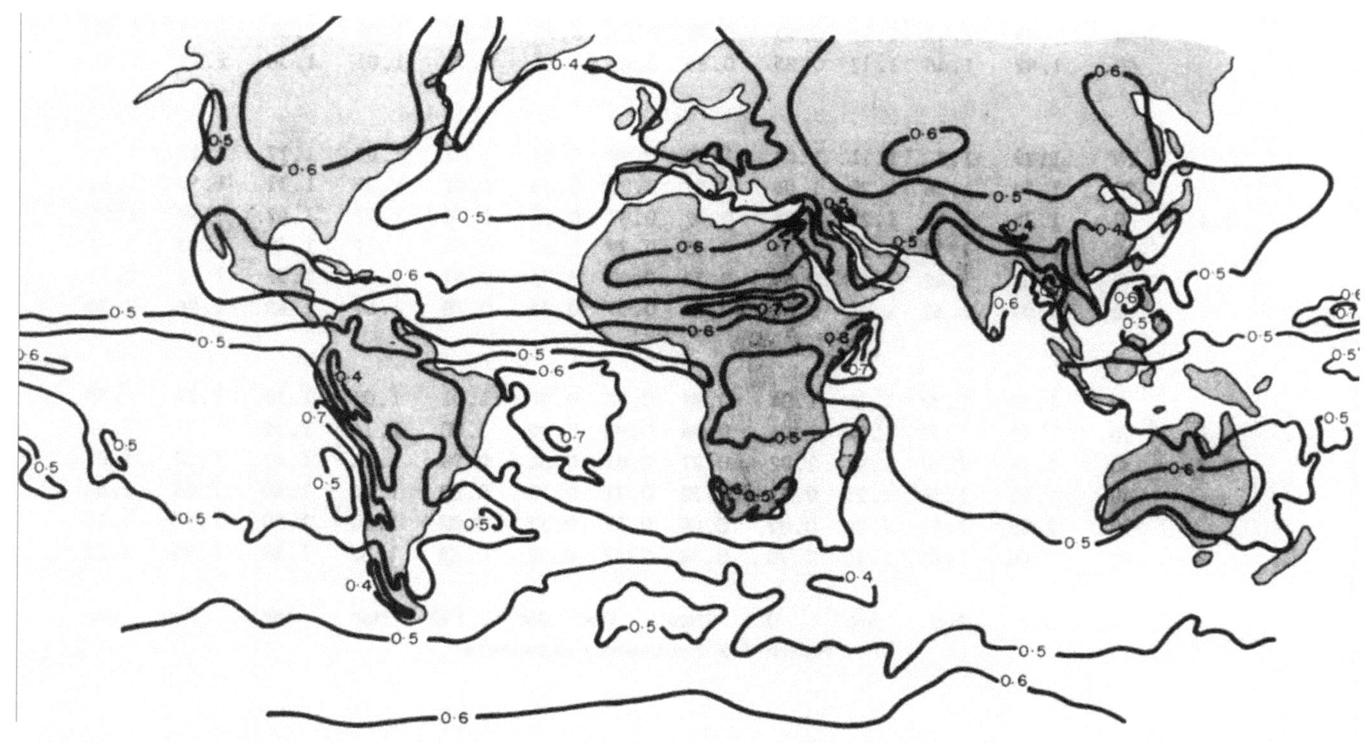

Figure A2. Clearness Index Map for February

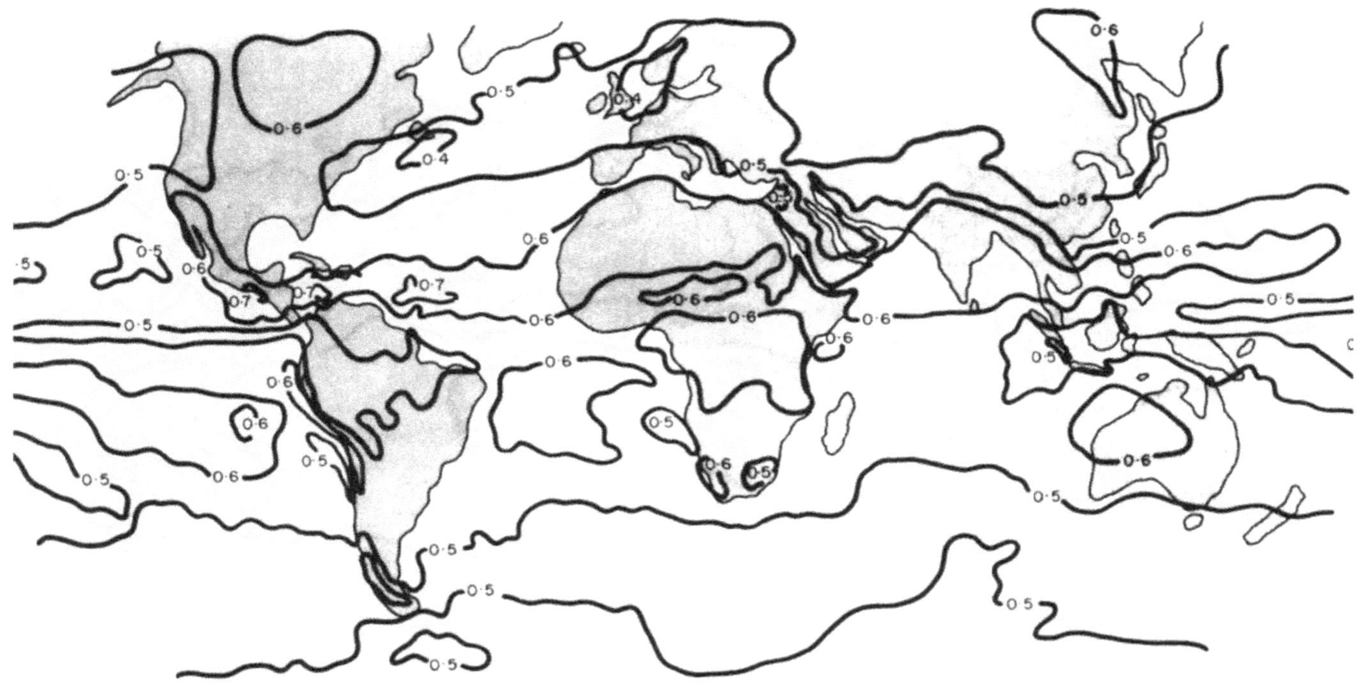

Figure A3. Clearness Index Map for March

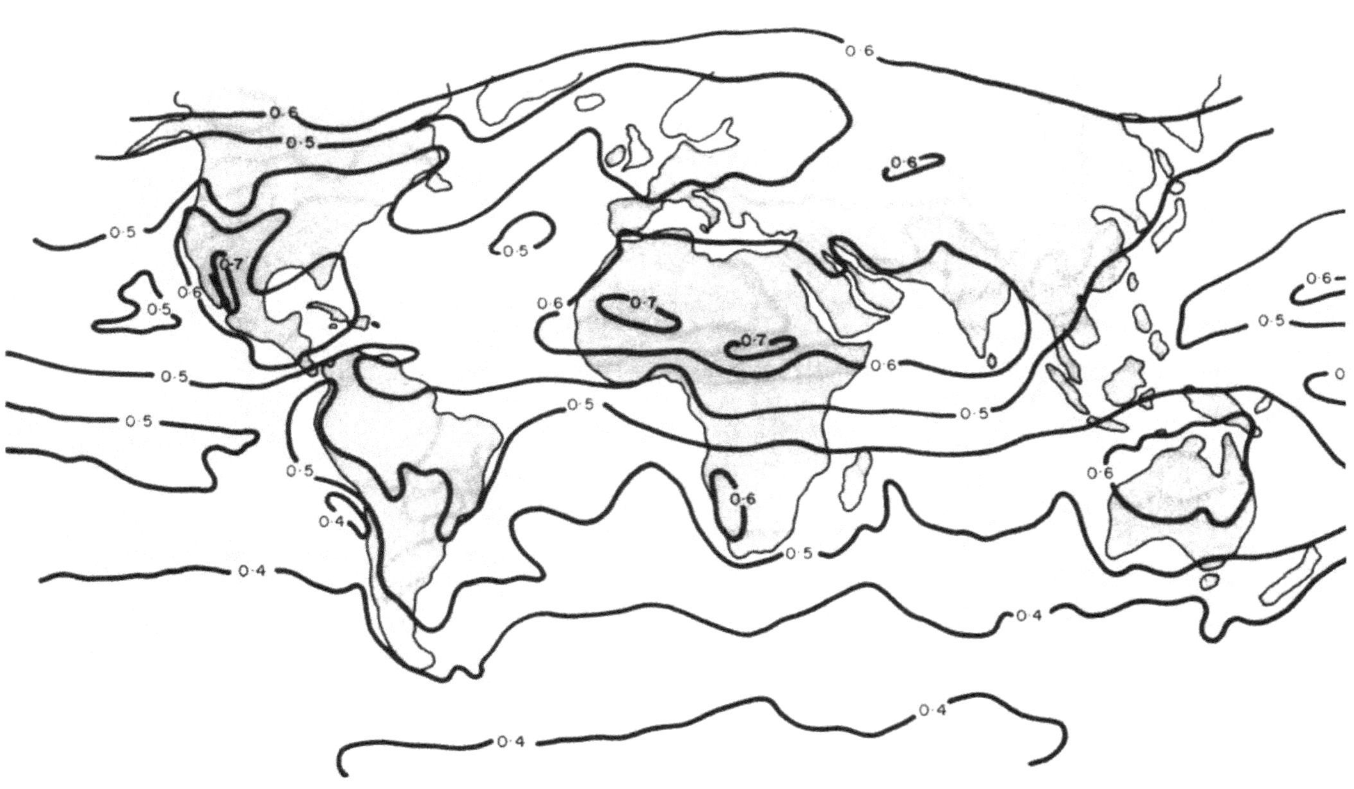

Figure A4. Clearness Index Map for April

Figure A5. Clearness Index Map for May

Figure A6. Clearness Index Map for June

Figure A7. Clearness Index Map for July

Figure A8. Clearness Index Map for August

Figure A9. Clearness Index Map for September

Figure A10. Clearness Index Map for October

Figure A11. Clearness Index Map for November

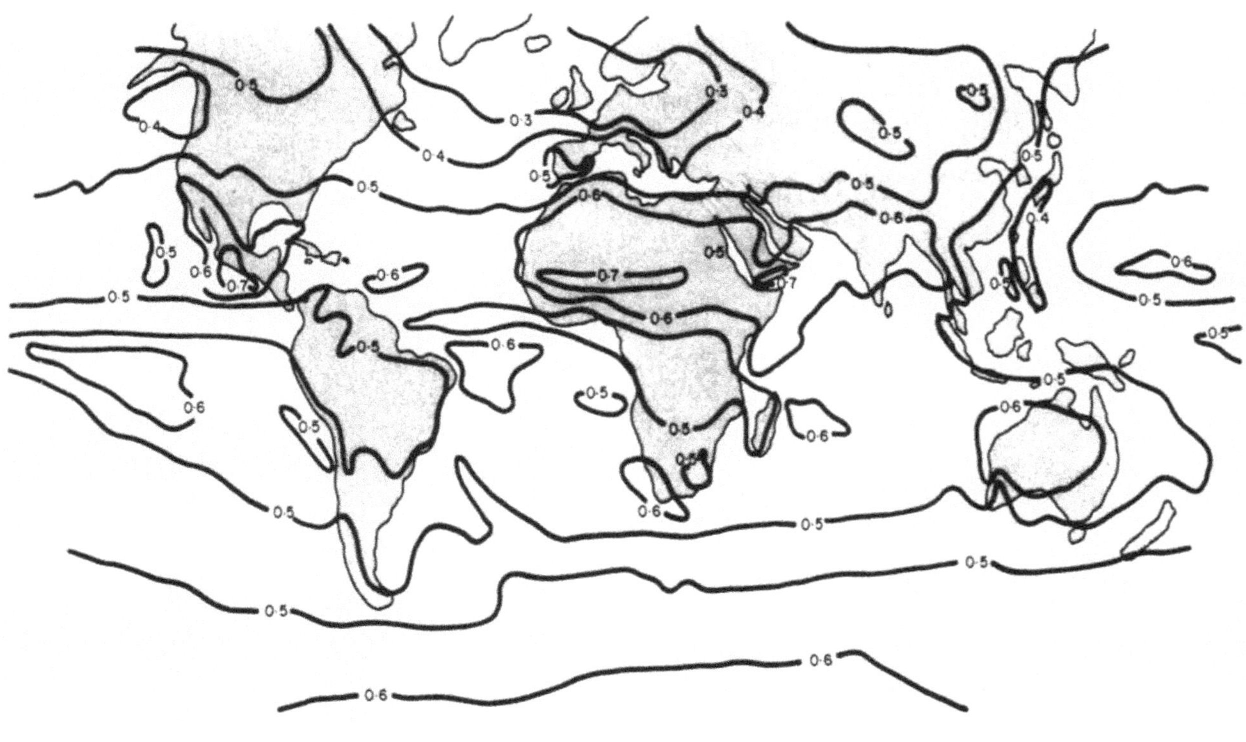

Figure A12. Clearness Index Map for December

<u>**APPENDIX 2.  CALCULATION OF THE PRESENT WORTH**</u>

For a future cost or benefit (Cr), payable in N years time, which is inflating at a fixed percentage "i" each year and discounted at a rate "d", the Present Worth is given by:

$$PW = Cr.Pr$$

$$\text{with} \quad Pr = \frac{(1 + i)^{N-1}}{(1 + d)^N}$$

For a payment or benefit Ca ($) occuring annually for a period of N years which is inflating at a rate i per year and discounted at a rate d, the Present Worth is

$$PW = Ca.Pa$$

$$\text{with} \quad Pa = \left[ 1 - \left(\frac{1 + i}{1 + d}\right)^N \right] / (d - i) \quad \text{for } i \neq d$$

$$\text{or} \quad Pa = N/(1 + i) \qquad \text{for } i = d$$

Tables A11 and A12 give the factor Pa, Pr for selected values of N, i and d.

| Discount<br>Rate (d) | Inflation<br>Rate (i) | Factor Pr for given number of year | | | | |
|---|---|---|---|---|---|---|
| | | 5 | 10 | 15 | 20 | 30 |
| 0.0 | 0.00 | 1.00 | 1.00 | 1.00 | 1.00 | 1.00 |
| | 0.05 | 1.21 | 1.55 | 1.98 | 2.53 | 4.12 |
| | 0.10 | 1.46 | 2.36 | 3.78 | 6.11 | 15.86 |
| | 0.15 | 1.75 | 3.53 | 7.07 | 14.23 | 57.57 |
| | 0.20 | 2.07 | 5.16 | 12.84 | 31.95 | 197.81 |
| 0.05 | 0.00 | 0.78 | 0.61 | 0.48 | 0.38 | 0.23 |
| | 0.05 | 0.95 | 0.95 | 0.95 | 0.95 | 0.95 |
| | 0.10 | 1.15 | 1.45 | 1.82 | 2.30 | 3.67 |
| | 0.15 | 1.37 | 2.16 | 3.40 | 5.36 | 13.37 |
| | 0.20 | 1.62 | 3.17 | 6.17 | 12.04 | 23.47 |
| 0.10 | 0.00 | 0.62 | 0.39 | 0.24 | 0.15 | 0.06 |
| | 0.05 | 0.75 | 0.60 | 0.47 | 0.37 | 0.23 |
| | 0.10 | 0.91 | 0.91 | 0.91 | 0.91 | 0.91 |
| | 0.15 | 1.08 | 1.36 | 1.69 | 2.11 | 3.30 |
| | 0.20 | 1.29 | 1.99 | 3.07 | 4.75 | 11.34 |
| 0.15 | 0.00 | 0.50 | 0.25 | 0.12 | 0.06 | 0.02 |
| | 0.05 | 0.60. | 0.38 | 0.24 | 0.15 | 0.08 |
| | 0.10 | 0.73 | 0.58 | 0.47 | 0.37 | 0.24 |
| | 0.15 | 0.87 | 0.87 | 0.87 | 0.87 | 0.87 |
| | 0.20 | 1.03 | 1.27 | 1.58 | 1.95 | 2.99 |
| 0.20 | 0.00 | 0.40 | 0.16 | 0.06 | 0.03 | 0.00 |
| | 0.05 | 0.49 | 0.25 | 0.13 | 0.06 | 0.02 |
| | 0.10 | 0.59 | 0.38 | 0.25 | 0.16 | 0.07 |
| | 0.15 | 0.70 | 0.57 | 0.46 | 0.37 | 0.24 |
| | 0.20 | 0.83 | 0.83 | 0.83 | 0.83 | 0.83 |

Table A11    Present Worth Factor Pr for selected values of inflation
rate, discount rate and number of year

| Discount<br>Rate (d) | Inflation<br>Rate (i) | Factor Pa for given number of year | | | | |
|---|---|---|---|---|---|---|
| | | 5 | 10 | 15 | 20 | 30 |
| 0.0 | 0.00 | 5.00 | 10.00 | 15.00 | 20.00 | 30.00 |
| | 0.05 | 5.52 | 12.58 | 21.58 | 33.07 | 66.00 |
| | 0.10 | 6.11 | 15.94 | 31.77 | 57.28 | 164.00 |
| | 0.15 | 6.74 | 20.30 | 47.58 | 102.44 | 435.00 |
| | 0.20 | 7.44 | 25.96 | 72.04 | 186.69 | 1182.00 |
| 0.05 | 0.00 | 4.32 | 7.72 | 10.38 | 12.46 | 15.37 |
| | 0.05 | 4.76 | 9.52 | 14.28 | 19.05 | 28.57 |
| | 0.10 | 5.23 | 11.85 | 20.19 | 30.71 | 60.75 |
| | 0.15 | 5.75 | 14.84 | 29.14 | 51.68 | 143.20 |
| | 0.20 | 6.33 | 18.67 | 42.74 | 89.66 | 359.49 |
| 0.10 | 0.00 | 3.79 | 6.14 | 7.61 | 8.51 | 9.42 |
| | 0.05 | 4.15 | 7.44 | 10.05 | 12.11 | 15.05 |
| | 0.01 | 4.54 | 9.09 | 13.64 | 18.18 | 27.27 |
| | 0.15 | 4.98 | 11.19 | 18.96 | 28.65 | 55.89 |
| | 0.20 | 5.45 | 13.87 | 26.88 | 46.99 | 126.04 |
| 0.15 | 0.00 | 3.35 | 5.02 | 5.85 | 6.26 | 6.56 |
| | 0.05 | 3.65 | 5.97 | 7.44 | 8.38 | 9.35 |
| | 0.10 | 3.98 | 7.18 | 9.73 | 11.78 | 14.73 |
| | 0.15 | 4.35 | 8.69 | 13.04 | 17.39 | 26.08 |
| | 0.20 | 4.74 | 10.61 | 17.87 | 26.85 | 51.70 |
| 0.20 | 0.00 | 2.99 | 4.19 | 4.67 | 4.87 | 4.98 |
| | 0.05 | 3.25 | 4.91 | 5.77 | 6.20 | 6.54 |
| | 0.10 | 3.52 | 5.31 | 7.29 | 8.24 | 9.26 |
| | 0.15 | 3.83 | 6.93 | 9.44 | 11.46 | 14.42 |
| | 0.20 | 4.16 | 8.33 | 12.50 | 16.67 | 25.00 |

Table  A12  Present Worth Factor Pa for selected values of inflation
rate, discount rate and number of years

APPENDIX 3.   EQUATIONS USED FOR ESTIMATING THE SIZE OF A PV ARRAY

A PV array is rated by its electrical output at a temperature of 25°C under a solar irradiance of 1000 W/m$^2$:

$$Wp = \eta_r \, A \, 1000 \qquad (1)$$

where $\quad$ Wp is the array rating in watts
$\eta_r$ is the PV array efficiency at the reference operating temperature (25°C)
A is the cell area in m$^2$

The cell area required to provide a daily energy output ($E_e$ MJ) for a daily solar irradiation (H MJ/m$^2$) incident on the array is:

$$A = E_e/(\eta_{pv}H) \qquad (2)$$

where $\eta_{pv}$ is the daily array efficiency under actual operating conditions. It is generally lower than $\eta_r$ because of temperature effects and the impedance mismatch between motor and array:

$$\eta_{pv} = Fm \, (1 - \beta(T_{cell} - 25)\eta_r \qquad (3)$$

The first term on the right hand side of Equation 3, (Fm) is the matching factor i.e. the ratio of electrical output under actual operating conditions to the electrical output if the array was operating at its maximum power point. The second term is the reduction in efficiency due to increases in cell temperature. $\beta$ is the cell temperature coefficient.

Substituting for $\eta_{pv}$ (from Equation 3), Equation (2) becomes:

$$A = E_e/(Fm(1 - \beta(T_{cell} - 25))\eta_r \, H) \qquad (4)$$

A is the cell area in m$^2$ required to provide a daily output of $E_e$ MJ. To determine the required array rating, Equation (4) is substituted into Equation (1):

$$Wp = 1000 \, E_e/(Fm(1 - \beta(T_{cell}-25)) \, H) \qquad (5)$$

Equation 5 has been used to derive the nomogram shown in Figure 19, using the following parameters: Fm = 0.9, $\beta$ = 0.005 per °C and $T_{cell}$ = 40°C, to give

$$Wp = 1200 \, E_e/H \qquad (6)$$

The electrical energy requirements ($E_e$) are related to the hydraulic energy requirements ($E_h$) by

$$E_e = E_h/\eta_{sub} \qquad (7)$$

where $\eta_{sub}$ is the daily energy efficiency of the sub-system.

For readers wishing to calculate the array rating for alternative cell temperatures the following approximate procedure can be used to estimate cell temperature:

For an average daily solar irradiation of H J/m$^2$ the energy <u>absorbed</u> as heat by the PV array is $\alpha$H, with $\alpha$ the array absorption coefficient. This energy <u>absorbed</u> must be balanced by heat losses from the array which can be approximated by $U(T_{cell} - T_a)\Delta t$:

$$\alpha H = U(T_{cell} - T_a)\Delta t \qquad (8)$$

or $\qquad T_{cell} = \dfrac{\alpha H}{U\Delta t} + T_a \qquad (9)$

where U is the array thermal loss coefficient (in W/m$^2$/°C)
$T_{cell}$ is the mean daily cell temperature
$T_a$ is the mean daylight hours ambient air temperature
and $\Delta t$ is the number of daylight seconds.

As a broad approximation the following values can be assumed:

$\alpha$ = 0.8
U = 25 W/m$^2$/ °C
$\Delta t$ = 12 hours = 43,200 seconds

If H is recorded in MJ/m$^2$, the daily average cell temperature becomes:

$$T_{cell} = 0.75 \, H + T_a \qquad (10)$$

97

<u>**APPENDIX 4.  EXAMPLES**</u>

<u>Example 1.  Village Water Supply</u>

What  is  the  unit water cost for (a) a solar pump with a 1 day storage tank and  (b)  handpumps,
located in a village of 500 people at Tiruchirapalli India (latitude 10°N, longitude 78°E).

Use the following cost and technical and economic data:

|  |  |  |
|---|---|---|
| Per capita water consumption | : | 40 litres / day |
| Period of analysis | : | 20 years |
| Discount rate | : | 10% |
| Storage tank height | : | 2 m |
| Borehole depth | : | 20 m |
| PV array tilt | : | 30 degrees |
| Borehole cost | : | $10 per metre |
| Solar pump costs | : | as Tables 12 and 13 |
| Storage tank cost | : | $10 per m³ |
| Motor/pump lifetime | : | 7.5 years |
| PV array lifetime | : | 15 years |
| Handpump lifetime | : | 10 years |
| Handpump cost | : | $200 |
| Handpump maintenance | : | $50 per pump per year |

(a)  Solar pump

The  four formats shown in Tables A13 to A16 have been completed to calculate the unit  water
cost for the solar pump:

1.   Calculation of hydraulic energy requirement (Table A13)

For a per capita consumption of 40 litres per day the daily water requirement is
40 x 500 = 20,000 litres or 20 m³.  The total static lift is the sum of the borehole depth and
the storage tank height i.e. 22 metres.  The daily hydraulic energy requirement is calculated
from the equation given in section 1.2:

Hydraulic energy requirement = $\dfrac{20 \times 9.81 \times 22}{1000}$ = 4.32 MJ

The energy requirements are identical each month.

2.   Calculation of available solar energy (Table A14)

Table A1 gives the month by month extra-terrestrial solar irradiation for latitude 10°N.
The  clearness  index for each month is obtained from Figures A1 to A12 and the  monthly
global irradiation is given by

(clearness index x extra-terrestrial irradiation)

The tilt factors for a latitude 10° N are given in Table A4 and the irradiation incident
on the PV array is calculated from

(Tilt factor x horizontal irradiation)

3.   Solar pump system sizing (Table A15)

The design month is either June or July. Using data for the design month the required PV
array  rating  is  obtained  from  Figure 19,  and the  pump  size  from  Figure 20.

4.   Performance specification (Table A16)

The  performance  specification of the solar pump is completed and the  month  by  month
water  output is calculated as outlined in section 3.4.   This shows the overcapacity of
the pump in the months of high solar irradiation.

5.   Unit water cost  (Table A17)

Using the cost data from Tables 12 and 13,  the capital and recurrent costs are  calculated as detailed in Table A17.  The unit water cost for the solar pump is 22.3 cents per
m³.

(b) Handpump

Table A18 shows the unit water cost format completed for the handpump. The total lift is taken to be 21 metres (allowing for an extra 1m for the handpump itself). The number of handpumps for the village is determined from figure 25. For a 'four hour day' the number of people required for 20 m³ per day at a 21 m lift is outside the range of the figure. However, the number of pumps required can be deduced by doubling the number needed to pump 10m³. From figure 25, it can be seen that a water consumption of 10 m³ per day at a 21 metre lift requires 3 people. Hence 6 people would be occupied for a period of 4 hours per day to provide the total water requirement for the village and it can be expected that up to 6 separate handpumps would be required.

The capital cost of 6 handpumps and 6 boreholes is $2400. Six replacements would be required in 10 years time at a Present Worth of $462 and the annual maintenance costs would be $300 The resulting unit water cost is 6.2 cents per m³. Hence the handpump is the cheaper option for this application.


## Example 2.  Livestock Water Supply

What is the unit water cost for (a) a solar pump with a 1 day store and (b) a windpump with a 1 day store supplying water for a herd of 375 cattle in Nanyuki, Kenya (latitude 0° , longitude 37°E).

Use the following cost, technical and economical data:

| | | |
|---|---|---|
| Per capita water consumption | : | 40 litres / day |
| Period of analysis | : | 20 years |
| Discount rate | : | 10% |
| Storage tank height | : | 1 m |
| Borehole depth | : | 29 m |
| PV array tilt | : | 10 |
| Borehole cost | : | $50 per metre |
| Solar pump costs | : | as Tables 12 and 13 |
| Storage tank cost | : | $40 per m³ |
| Motor/pump liftime | : | 7.5 years |
| PV array lifetime | : | 15 years |
| Windpump structural lifetime | : | 30 years |
| Pump lifetime | : | 10 years |
| Windpump costs | : | as Table 14 |
| Design month windspeed | : | estimated to lie between 2 m/s and 2.5 m/s |

(a) Solar pump

Tables A20 to A24 show the completed formats for the solar pump. The unit water cost is 31.4 cents per m³.

(b) Windpump

The windpump rotor size for a water requirement of 15 m³ per day at a 30 metre static lift is obtained from figure 23. At a design month wind speed of 2 m/s the output of a windpump for a 30 m static lift is approximately 0.25 m³ per m² of rotor. Similarly at a wind speed of 2.5 m/s the output is approximately 0.5 m³ per m² of rotor. Hence to provide 15 m³ of water per day windpumps of rotor areas equal to 60 m² and 30 m² would be required for mean wind speeds of 2.0 and 2.4 m/s respectively. From Table 14 the average cost of a windpump is $300 per m² . This gives a capital cost between $9000 and $18000 for this particular application.

Table A24 and A25 show the completed format sheets of the windpump. The unit water cost at this location would be between 23.5 and 41.7 cents per m³. Hence the solar pump would be cheaper if the design month windspeed was 2 m/s but more expensive if the design month wind speed was 2.5 m/s.

HYDRAULIC ENERGY REQUIREMENTS

Location ..TIRUCHI.......   Latitude ....10°N.......

Delivery pipe head loss ..... %;   Delivery pipe length .... m

| Month | Pumped volume requirement (m³/day) | Static head (m) | Dynamic head loss (m) | Total head (m) | Hydraulic energy (MJ/day) |
|---|---|---|---|---|---|
| Jan | 20 | 22 | - | 22 | 4·32 |
| Feb | 20 | 22 | - | 22 | 4·32 |
| March | 20 | 22 | - | 22 | 4·32 |
| April | 20 | 22 | - | 22 | 4·32 |
| May | 20 | 22 | - | 22 | 4·32 |
| June | 20 | 22 | - | 22 | 4·32 |
| July | 20 | 22 | - | 22 | 4·32 |
| Aug | 20 | 22 | - | 22 | 4·32 |
| Sept | 20 | 22 | - | 22 | 4·32 |
| Oct | 20 | 22 | - | 22 | 4·32 |
| Nov | 20 | 22 | - | 22 | 4·32 |
| Dec | 20 | 22 | - | 22 | 4·32 |

Figure A13. Hydraulic energy requirements at Tiruchirapalli

SOLAR ENERGY AVAILABILITY

Location TIRUCHI.   Latitude ..10°N...   Array Tilt .30°.....

Longitude ..78°E...

| Month | Extra-terrestrial Irradiation MJ/m² | Clearness Index | Global Horizontal Irradiation MJ/m² | Tilt Factor | Global Irradiation on Array MJ/m² |
|---|---|---|---|---|---|
| Jan | 32 | 0·7 | 22·4 | 1·34 | 30 |
| Feb | 34 | 0·6 | 20·4 | 1·14 | 23 |
| March | 37 | 0·6 | 22·2 | 0·99 | 22 |
| April | 38 | 0·6 | 22·8 | 0·85 | 19 |
| May | 37 | 0·6 | 22·2 | 0·74 | 16 |
| June | 37 | 0·5 | 18·5 | 0·74 | 14 |
| July | 37 | 0·5 | 18·5 | 0·76 | 14 |
| Aug | 37 | 0·5 | 18·5 | 0·83 | 15 |
| Sept | 37 | 0·6 | 22·0 | 0·94 | 21 |
| Oct | 35 | 0·5 | 17·5 | 1·07 | 19 |
| Nov | 32 | 0·5 | 16·0 | 1·18 | 19 |
| Dec | 31 | 0·6 | 18·6 | 1·31 | 24 |

Figure A14. Solar energy availability at Tiruchirapalli

## SOLAR PUMP SYSTEM SIZING

Location *TIRUCHI*. Latitude ..*10°N*.. Longitude ...*78°E*..

Design Month   *JUNE/JULY*

Design Month hydraulic energy requirement ......*4.32*..... MJ

Design Month head ............................*22*.... metres

Design Month global irradiation on array .......*14*..... MJ/m$^2$

Average sub system energy efficiency ...........*45*.......... %

Peak sub system power efficiency .................*45*......... %

| Step | Calculations |
|---|---|
| 1. PV array size | Required electrical energy ....*9.6*...... MJ<br>PV array size ...............*920*.......Wp |
| 2. Motor size | Rated motor input power ....*820*..... Watts |
| 3. Pump size | Rated peak hydraulic power ..*370*.... Watts<br>Rated flow rate ..............*1.7*..... lit/sec<br>at ..*22*... metres head |
| 4. Pipe Diameter | Diameter ...*—*..... (mm) for head loss of<br>...*—*..... (m) at rated peak flow rate |

Figure A15. Results of solar pump system sizing for Tiruchirapalli

## SOLAR PUMP PERFORMANCE SPECIFICATION

Location *TIRUCHI*. Latitude ..*10°N*... Longitude ...*78°E*...

1. Water source mean static lift ...................*20*...... m

2. Delivery system   Type ....................................
   Length .............................................. m
   Pipe diameter ....................................... mm
   Efficiency .......................................... %

3. Storage system (when applicable)
   Volume ......................*20*...... m$^3$
   Height ......................*2*....... m

4. Design month details
   End use water requirement ..*20*.. m$^3$/day
   Pumped water requirement ...*20*... m$^3$/day
   Hydraulic energy requirement*4.32* MJ/day
   Solar irradiation on
   PV array ...............*14*..... MJ/m$^2$/day

5. Solar pump performance and water requirement

|  | Jan | Feb | Mar | Apr | May | Jun | Jul | Aug | Sep | Oct | Nov | Dec |
|---|---|---|---|---|---|---|---|---|---|---|---|---|
| Solar Irradiation (MJ/m$^2$/day) on PV array | 30 | 23 | 22 | 19 | 16 | 14 | 14 | 15 | 21 | 19 | 19 | 24 |
| Pumped water requirement (m$^3$ per day) | 20 | 20 | 20 | 20 | 20 | 20 | 20 | 20 | 20 | 20 | 20 | 20 |
| Pumped water (m$^3$ output/day) | 43 | 33 | 31 | 27 | 23 | 20 | 20 | 21 | 30 | 27 | 27 | 34 |

6. Solar pump specification
   PV array size ...............*920*... Wp
   PV array tilt ................*30°*....
   Sub-system energy efficiency .*45*.... %
   Sub-system peak power efficiency .*45*. %
   Motor Rated Power ...........*820*.... W
   Pump rating *1.7* lit/sec at *22* m head

Figure A16. Solar pump performance specification for Tiruchirapalli

| UNIT WATER COST FOR A SMALL SCALE PUMPING SYSTEM |
| --- |

**1. System Description**

Location..TIRUCHI...  Latitude....10°N.....

Design Month..JUNE/JULY  System Head...22.....m

Design Month water requirement.20..m³/day  Power Source.SOLAR...

Annual water requirement..7300...m³  Size....820.Wp.....

**2. Cost Analysis**

Period of analysis....20......years  Discount Rate...10.%...

Pa.......851.............

|  | Annual Cost | Present Worth |
| --- | --- | --- |
| Capital Cost |  | $ 10650 |
| Replacements |  | $ 1704 |
| Maintenance | $ 80 | $ 681 |
| Operating | - | - |
| Total Annual Cost | $ 80 |  |
| Life Cycle Cost (LCC) |  | $ 13035 |
| Annual equivalent of LCC (ALCC) |  | $ 1531 |
| Unit Water Cost. | 21.0 cents/m³ | |

Table A17. Unit water cost calculations for a solar pump at Trichirapali

| UNIT WATER COST FOR A SMALL SCALE PUMPING SYSTEM |
|---|

1.  System Description

   Location..**TIRUCHI**..            Latitude....**10°N**.....

   Design Month...**-**......            System Head...**21**.....m

   Design Month water requirement **20**..m$^3$/day  Power Source.**HAND**...

   Annual water requirement..**7300**...m$^3$   Size.....**6 off**.......

2.  Cost Analysis

   Period of analysis...**20**......years        Discount Rate..**10%**....

   Pa......**8·51**............

|  | Annual Cost | Present Worth |
|---|---|---|
| Capital Cost |  | **$ 2400** |
| Replacements |  | **$ 462** |
| Maintenance | **$ 150** | **$ 1276** |
| Operating | **-** | **-** |
| Total Annual Cost | **$ 150** |  |
| Life Cycle Cost (LCC) |  | **4138** |
| Annual equivalent of LCC (ALCC) |  | **486** |
| Unit Water Cost. | **6·7 cents/m$^3$** ||

Table A18. Unit water cost calculations for 6 handpumps at Trichurapali

HYDRAULIC ENERGY REQUIREMENTS

Location ..NANYUKI.......          Latitude .......0°........

Delivery pipe head loss ..:.. %;   Delivery pipe length ..:.. m

| Month | Pumped volume requirement (m$^3$/day) | Static head (m) | Dynamic head loss (m) | Total head (m) | Hydraulic energy (MJ/day) |
|---|---|---|---|---|---|
| Jan | 15 | 30 | - | 30 | 4.41 |
| Feb | 15 | 30 | - | 30 | 4.41 |
| March | 15 | 30 | - | 30 | 4.41 |
| April | 15 | 30 | - | 30 | 4.41 |
| May | 15 | 30 | - | 30 | 4.41 |
| June | 15 | 30 | - | 30 | 4.41 |
| July | 15 | 30 | - | 30 | 4.41 |
| Aug | 15 | 30 | - | 30 | 4.41 |
| Sept | 15 | 30 | - | 30 | 4.41 |
| Oct | 15 | 30 | - | 30 | 4.41 |
| Nov | 15 | 30 | - | 30 | 4.41 |
| Dec | 15 | 30 | - | 30 | 4.41 |

Figure A19. Hydraulic energy requirements at Nanyuki

SOLAR ENERGY AVAILABILITY

Location NANYUKI  Latitude ...0°...... Array Tilt ..10°.....

Longitude ..37°E...

| Month | Extra-terrestrial Irradiation MJ/m$^2$ | Clearness Index | Global Horizontal Irradiation MJ/m$^2$ | Tilt Factor | Global Irradiation on Array MJ/m$^2$ |
|---|---|---|---|---|---|
| Jan | 38 | 0.5 | 18.0 | 1.08 | 19 |
| Feb | 37 | 0.6 | 22.2 | 1.04 | 23 |
| March | 38 | 0.5 | 19.0 | 1.00 | 19 |
| April | 36 | 0.6 | 21.6 | 0.95 | 20 |
| May | 34 | 0.5 | 17.0 | 0.92 | 16 |
| June | 33 | 0.5 | 16.5 | 0.90 | 15 |
| July | 34 | 0.5 | 17.0 | 0.91 | 15 |
| Aug | 35 | 0.5 | 17.5 | 0.94 | 16 |
| Sept | 37 | 0.5 | 18.5 | 0.98 | 18 |
| Oct | 37 | 0.5 | 18.5 | 1.02 | 19 |
| Nov | 36 | 0.5 | 18.0 | 1.06 | 19 |
| Dec | 35 | 0.5 | 17.5 | 1.07 | 18 |

Figure A20. Solar energy availability at Nanyuki

## SOLAR PUMP SYSTEM SIZING

Location **NANYUKI**  Latitude ...0°.....  Longitude ..37°E...

Design Month  **JUNE / JULY**

Design Month hydraulic energy requirement ......4.41..... MJ

Design Month head ...............................30.... metres

Design Month global irradiation on array .........15..... MJ/m$^2$

Average sub system energy efficiency ..................45.. %

Peak sub system power efficiency ......................45.. %

| Step | Calculations |
|---|---|
| 1.  PV array size | Required electrical energy ....9.80... MJ<br><br>PV array size ..................780....Wp |
| 2.  Motor size | Rated motor input power ........780... Watts |
| 3.  Pump size | Rated peak hydraulic power ....350. Watts<br><br>Rated flow rate ...............1.2.. lit/sec<br><br>at .30... metres head |
| 4.  Pipe Diameter | Diameter ....–..... (mm) for head loss of<br>....–.... (m) at rated peak flow rate |

Figure A21. Results of solar pump system sizing for Nanyuki

## SOLAR PUMP PERFORMANCE SPECIFICATION

Location **NANYUKI**  Latitude ...0°.....  Longitude ..37°E...

1.  Water source mean static lift ...................29..... m

2.  Delivery system    Type .........................–....<br>    Length .............................. m<br>    Pipe diameter ....................... mm<br>    Efficiency .......................... %

3.  Storage system (when applicable)<br>    Volume ..................15.......... m$^3$<br>    Height .................1........... m

4.  Design month details<br>    End use water requirement ..15.... m$^3$/day<br>    Pumped water requirement ...15... m$^3$/day<br>    Hydraulic energy requirement 4.41 MJ/day<br>    Solar irradiation on<br>    PV array ...............15.... MJ/m$^2$/day

5.  Solar pump performance and water requirement

|  | Jan | Feb | Mar | Apr | May | Jun | Jul | Aug | Sep | Oct | Nov | Dec |
|---|---|---|---|---|---|---|---|---|---|---|---|---|
| Solar Irradiation (MJ/m$^2$/day) on PV array | 19 | 23 | 19 | 20 | 16 | 15 | 15 | 16 | 18 | 19 | 19 | 18 |
| Pumped water requirement (m$^3$ per day) | 15 | 15 | 15 | 15 | 15 | 15 | 15 | 15 | 15 | 15 | 15 | 15 |
| Pumped water (m$^3$ output/day) | 19 | 23 | 19 | 20 | 16 | 15 | 15 | 16 | 18 | 19 | 19 | 18 |

6.  Solar pump specification<br>    PV array size ...............780.. Wp<br>    PV array tilt ................10°.....<br>    Sub-system energy efficiency ..45... %<br>    Sub-system peak power efficiency 45.. %<br>    Motor Rated Power ............780.. W<br>    Pump rating .1.2 lit/sec at 30. m head

Figure A22. Solar pump performance specification for Nanyuki

| UNIT WATER COST FOR A SMALL SCALE PUMPING SYSTEM |
|---|

**1. System Description**

Location.. **NANYUKI** .                    Latitude.... **0°** ..........

Design Month.. **JUNE/JULY** ..           System Head... **30** .....m

Design Month water requirement. **15** .m³/day  Power Source.. **SOLAR** ..

Annual water requirement. **5475** ....m³   Size.... **780 Wp** .....

**2. Cost Analysis**

Period of analysis.... **20** .....years     Discount Rate.. **10%** ....

Pa...... **8·51** ...........

|  | Annual Cost | Present Worth |
|---|---|---|
| Capital Cost |  | **$ 11850** |
| Replacements |  | **$ 1504** |
| Maintenance | **$ 80** | **$ 681** |
| Operating | - | - |
| Total Annual Cost | **$ 80** |  |
| Life Cycle Cost (LCC) |  | **14035** |
| Annual equivalent of LCC (ALCC) |  | **1649** |
| Unit Water Cost. | **30·1 cents/m³** | |

Table A23. Unit water cost calculations for a solar pump at Nanyuki

UNIT WATER COST FOR A SMALL SCALE PUMPING SYSTEM

1.  System Description

Location. **NANYUKI**.                          Latitude.... $0°$ ........

Design Month. **FEB**....                       System Head.. **30** ....m

Design Month water requirement. **15** .m$^3$/day  Power Source. **WIND** ....

Annual water requirement. **5475** ....m$^3$       Size.... **60 m$^2$** ........

2.  Cost Analysis

Period of analysis.... **20** ......years        Discount Rate.. **10 %** ...

Pa..... **8·51** .............

| | Annual Cost | Present Worth |
|---|---|---|
| Capital Cost | | **$ 20,100** |
| Replacements | | **$ 680** |
| Maintenance | **$ 68** | **$ 579** |
| Operating | - | - |
| Total Annual Cost | **$ 68** | |
| Life Cycle Cost (LCC) | | **$ 21359** |
| Annual equivalent of LCC (ALCC) | | **$ 2510** |
| Unit Water Cost. | **45·8 cents/m$^3$** | |

Figure A24. Unit water cost calculations for a windpump at Nanyuki (a) wind speed = 2 m/s

UNIT WATER COST FOR A SMALL SCALE PUMPING SYSTEM

1.  System Description

    Location....**NANYUKI**....          Latitude.....**0°**........

    Design Month....**FEB**....        System Head...**30**.....m

    Design Month water requirement..**15**..m³/day Power Source..**WIND**....

    Annual water requirement...**5475**....m³   Size.....**30 m²**.......

2.  Cost Analysis

    Period of analysis.....**20**......years    Discount Rate...**10%**...

    Pa......**8.51**............

|  | Annual Cost | Present Worth |
|---|---|---|
| Capital Cost |  | **$ 11,100** |
| Replacements |  | **$ 340** |
| Maintenance | **$ 68** | **$ 579** |
| Operating | **-** | **-** |
| Total Annual Cost | **$ 68** |  |
| Life Cycle Cost (LCC) |  | **$ 12019** |
| Annual equivalent of LCC (ALCC) |  | **$ 1412** |
| Unit Water Cost. | **25.8 cents /m³** ||

Figure A25. Unit water cost calculations for a windpump at Nanyuki (b) wind speed = 2.5 m/s

<u>**APPENDIX 5. BLANK FORMAT SHEETS**</u>


Contents


Format Sheet for Calculation of Hydraulic Energy Requirements

Format Sheet for Calculation of Solar Energy Availabilty

Format Sheet for Calculation of System Size

Format Sheet for Specification of Solar Pump Perfromance

Format Sheet for Calculation of Unit Water Cost

## HYDRAULIC ENERGY REQUIREMENTS

Location .................    Latitude ................

Delivery pipe head loss .... %;  Delivery pipe length .... m

| Month | Pumped volume requirement ($m^3$/day) | Static head (m) | Dynamic head loss (m) | Total head (m) | Hydraulic energy (MJ/day) |
|-------|-----|-----|-----|-----|-----|
| Jan | | | | | |
| Feb | | | | | |
| March | | | | | |
| April | | | | | |
| May | | | | | |
| June | | | | | |
| July | | | | | |
| Aug | | | | | |
| Sept | | | | | |
| Oct | | | | | |
| Nov | | | | | |
| Dec | | | | | |

Format Sheet for Calculation of Hydraulic Energy Requirements

| SOLAR ENERGY AVAILABILITY | | | | | |
|---|---|---|---|---|---|
| Location .......... Latitude ........... Array Tilt .......... | | | | | |
| Longitude ......... | | | | | |
| Month | Extra-terrestrial Irradiation MJ/m$^2$ | Clearness Index | Global Horizontal Irradiation MJ/m$^2$ | Tilt Factor | Global Irradiation on Array MJ/m$^2$ |
| Jan | | | | | |
| Feb | | | | | |
| March | | | | | |
| April | | | | | |
| May | | | | | |
| June | | | | | |
| July | | | | | |
| Aug | | | | | |
| Sept | | | | | |
| Oct | | | | | |
| Nov | | | | | |
| Dec | | | | | |

Format Sheet for Calculation of Solar Energy Availabilty

<table>
<tr><td colspan="2" align="center">SOLAR PUMP SYSTEM SIZING</td></tr>
<tr><td colspan="2">

Location .......... Latitude .......... Longitude ..........

Design Month

Design Month hydraulic energy requirement ................. MJ

Design Month head .................................... metres

Design Month global irradiation on array .............. $MJ/m^2$

Average sub system energy efficiency ....................... %

Peak sub system power efficiency .......................... %

</td></tr>
<tr><td align="center">Step</td><td align="center">Calculations</td></tr>
<tr><td>

1. PV array size</td><td>

Required electrical energy ............. MJ

PV array size ...........................Wp

</td></tr>
<tr><td>

2.   Motor size</td><td>

Rated motor input power ............. Watts

</td></tr>
<tr><td>

3.   Pump size</td><td>

Rated peak hydraulic power .......... Watts

Rated flow rate .................... lit/sec

at ....... metres head

</td></tr>
<tr><td>

4.   Pipe Diameter</td><td>

Diameter ......... (mm) for head loss of
.......... (m) at rated peak flow rate

</td></tr>
</table>

Format Sheet for Calculation of System Size

| SOLAR PUMP PERFORMANCE SPECIFICATION |
| :---: |

Location ......... Latitude .......... Longitude ..........

1.  Water source mean static lift ........................... m

2.  Delivery system    Type ...................................
                       Length ............................... m
                       Pipe diameter ...................... mm
                       Efficiency ........................... %

3.  Storage system (when applicable)
                       Volume ............................. $m^3$
                       Height ............................... m

4.  Design month details
                       End use water requirement ....... $m^3$/day
                       Pumped water requirement ........ $m^3$/day
                       Hydraulic energy requirement .... MJ/day
                       Solar irradiation on
                       PV array .................... MJ/$m^2$/day

5.  Solar pump performance and water requirement

|  | Jan | Feb | Mar | Apr | May | Jun | Jul | Aug | Sep | Oct | Nov | Dec |
| --- | --- | --- | --- | --- | --- | --- | --- | --- | --- | --- | --- | --- |
| Solar Irradiation (MJ/$m^2$/day) on PV array |  |  |  |  |  |  |  |  |  |  |  |  |
| Pumped water requirement ($m^3$ per day) |  |  |  |  |  |  |  |  |  |  |  |  |
| Pumped water ($m^3$ output/day) |  |  |  |  |  |  |  |  |  |  |  |  |

6.  Solar pump specification
                       PV array size ....................... Wp
                       PV array tilt ..........................
                       Sub-system energy efficiency ......... %
                       Sub-system peak power efficiency ..... %
                       Motor Rated Power ................... W
                       Pump rating .... lit/sec at .... m  head

Format Sheet for Specification of Solar Pump Perfromance

| UNIT WATER COST FOR A SMALL SCALE PUMPING SYSTEM |

1. System Description

Location............. Latitude..............

Design Month......... System Head..........m

Design Month water requirement.....m$^3$/day Power Source..........

Annual water requirement..........m$^3$ Size..................

2. Cost Analysis

Period of analysis............years Discount Rate..........

Pa......................

|  | Annual Cost | Present Worth |
|---|---|---|
| Capital Cost |  |  |
| Replacements |  |  |
| Maintenance |  |  |
| Operating |  |  |
| Total Annual Cost |  |  |
| Life Cycle Cost (LCC) |  |  |
| Annual equivalent of LCC (ALCC) |  |  |
| Unit Water Cost. |  |  |

**Format Sheet for Calculation of Unit Water Cost**

<u>APPENDIX 6: EXAMPLE TENDER DOCUMENTS</u>

<u>FOR THE PROCUREMENT OF PV SOLAR PUMPS</u>


Contents


1.    Instructions to Tenderer

2.    System specification

3.    Questionnaire for Tenderers

4.    Price and delivery


Notes for Purchaser

The following tender documents are given for guidance only and may need modification for a particular purchaser's needs. The items listed below should must be completed by the purchaser prior to issue of the document:

o  Section 1.1    Complete dates for tender procedure.
o  Section 2.1    Specify application and location of system.
o  Section 2.2    Specify maximum sizes for containers in the shipment.
o  Section 2.3    Insert environmental conditions for application.
o  Section 2.6    Details of the site and application to be provided. Table 1 should specify the
                  month by month water requirement, and Figure 1 should indicate the layout of
                  the proposed site for the system.
o  Section 2.9    Specify the language in which installation instructions are to be submitted.

<u>1.</u>    <u>INSTRUCTIONS TO TENDERERS</u>

<u>1.1    Tender Procedure</u>

Tenders are required for a complete solar powered pumping system as described in the specification. The schedule for the purchase of system is as follows:

Tender forms issued by        ..........(day)............(month)........(year)

Completed tenders to be
returned by                   ..........(day)............(month)........(year)

Tenders awarded by            ..........(day)............(month)........(year)

Systems to be delivered by ......................................

The original tender shall be in the ............. language and shall be filled out in  ink or typewritten and will be made a part of the awarded contract.

<u>1.2    Adjudication Process</u>

Tenders will be primarily considered for:

    (i)    Performance
    (ii)   Durability
    (iii)  Cost effectiveness

The purchaser will not be bound to award a contract to the lowest, or any Tenderer.


<u>2.</u>    <u>SPECIFICATION</u>

<u>2.1    Scope</u>

This Specification is for the design,  manufacture, supply and delivery of a complete self-contained  solar  photovoltaic powered water pumping system suitable for  irrigation*/water supply* use in ..........................................................................

(* delete as apropriate)

The system to be supplied shall include:

o  Photovoltaic modules and array support structure

o  Motor

o  Pump

o  Pipework

o  All control equipment and wiring

o  All fixings and ancillaries necessary for complete construction and commissioning

o  Tools needed for assembly and maintenance

o  Spare parts

o  Documentation

The  clauses which follow outline the design parameters and other requirements to be observed for the fulfilment of the Contract.

<u>2.2    Design</u>

The  complete  system shall be robust, ·and capable of withstanding hard usage in  a  harsh environment. It  shall be resistant to damage from accidental misuse and reasonably resistant to vandalism and the attentions of animals, wild or domestic.

The  system shall be designed for assembly,  operation and servicing by unskilled personnel under  the guidance of a trained technician. The requirement for special tools or  instruments  to  install  and maintain the system shall be minimised and  all  tools  needed  for installation shall be supplied with the system. Foundations or other preparatory work shall be as simple as practicable.

The system shall be designed for assembly from units which can be packed in containers small enough to be easily man-handled and transported on small vehicles. The maximum permited dimensions for any one unit are: .............................

The system shall be designed to operate for a long lifetime with minimum deterioration of performance. The design life of the whole system shall be at least ten years with a minimal need for replacement of moving parts and wearing components such as motor brushes or water seals. Routine maintenance shall be minimised and maintenance work necessary shall be as simple as possible, requiring only a few basic tools for its execution.

## 2.3    Environmental Conditions

The system shall be designed to meet the requirements of this. Specification under the following environmental conditions:

(i)     Ambient air temperature between .... and .... (e.g: $-5^{0}$C and $45^{0}$C).

(ii)    Relative humidity up to .... at an ambient temperature of .... (e.g: 90% and $45^{0}$C).

(iii)   Wind speed up to .... km/hr (e.g: 150 km/hr) (for fixed installations).

(iv)    Water source temperature up to ....$^{0}$C (e.g: $35^{0}$C).

(v)     Water containing particles not exceeding ....mm (e.g: 0.3 mm).

(vi)    A maximum altitude above sea level of ....m (e.g: 2000 m).

The system should also be resistant to the following extremes of environment:
e.g:

(i)     Sand storms

(ii)    A high dissolved solid content in the pumped water

(iii)   A high sediment load in the pumped water

(iv)    Typhoon or hurricane winds

(v)     Overnight freezing temperatures

(vi)    .........................

The Contractor shall state the limits of environmental conditions under which the system is designed to operate.

## 2.4    Materials and Workmanship

All materials used shall be of first class quality in accordance with relevant national standards, carefully selected for the duty required, with particular regard given to resistance against corrosion and long term degradation. Workmanship and general finish shall be in accordance with the best modern practice.

## 2.5    Standards

Photovoltaic modules shall comply with the test requirements of either the current Photovoltaic Module Control Test specifications of the Commission of European Communities Joint Research Centre (Ispra Establishment), or the current Jet Propulsion Laboratory (California, USA) Module Test Specification.

## 2.6    Performance Requirement

### 2.6.1  Location

The pumping system to be supplied by the contractor is to be located as detailed below:

o   Name of nearest village/town:
o   Country:
o   Latitude:
o   Longitude:
o   Water source (river/canal/open well/borehole):

2.6.2  Required Performance

The required performance of the system is summarized in Table 1, along with the typical environmental conditions for the location. The system should provide average daily outputs as specified in Table 1 for each month, provided that the specified monthly mean average daily solar irradiation for the month is met or exceeded. The tenderer shall complete Table 1 using performance data for the system offered.

2.6.3  Installation Details

A sketch of the site is shown in Figure 1. The total static head at the site is detailed in Table 1 and includes ....... metres head for discharge above ground level (see Figure 1). The well/borehole details are (when applicable):

     (i)   diameter ....... m
     (ii)  lining/casing depth ....... m
     (iii) drawdown ....... m  at ....... 1/sec

## 2.7    Spare Parts

The Contractor shall supply with the system sufficient consumable items such as motor brushes and water seals which may need replacement to last for 10000 hours of operation. Spare nuts, bolts, washers etc. likely to be lost during shipment and erection shall also be supplied at the time of shipment.

## 2.8    Packing for Shipment

All equipment shall be carefully and suitably packed for the specific means of transportation to be used, so that it is protected against all weather and other conditions to which it may become subject.

Before despatch all equipment is to be thoroughly dried and cleaned internally. All external unpainted ferrous parts and machined surfaces shall be protected by an approved proprietary preservative, all openings shall be covered and all screwed connections plugged unless otherwise agreed.

Where moisture absorbants have been used for protection from corrosion during storage or transit, adequate information of their location and warning as to their removal shall be clearly indicated.

Paraffin-treated paper strips, plastic balls or similar may be used as filler material in boxes. Wood, wool, hay or straw as filler and for stiffening of the goods should not be used.

Complete assembly and operating instructions are to be included in packing.

## 2.9    Documentation

Prior to shipment of the equipment, the Contractor shall submit to the Purchaser the following documents: (Copies also should be shipped with system.)

   (i)   A list of components and assemblies to be shipped including all spare parts and tools
   (ii)  The size, weight and packing list for each package in the shipment.
  (iii)  Assembly instructions
   (iv)  Operating instructions
   (v)  Instructions for all maintenance operations and the schedule for any routine maintenance requirements
   (vi)  Sufficient descriptions of spare parts and components to permit identification for ordering replacements
  (vii)  Revised drawings of the equipment as built if different from the approved proposals

All documents shall be in the ........... language.

## 2.10   Tools

The Contractor shall provide with the pumping system two sets of any special tools and other equipment that are required for erecting, operating, maintaining and repairing the equipment. Special tools shall include such items as Allen or socket keys, box spanners, feeler gauges, grease guns etc. A single set of all other tools required for erection shall also be supplied.

2.11    Technical Support after Shipment

The Contractor shall be prepared to provide advice during the installation and warranty periods of the equipment supplied under the Contract. For this purpose he shall nominate a member of his executive or technical staff who may be contacted during normal office hours.

2.12    Insurance

The Contractor shall arrange for the equipment to be comprehensively insured for its full from the time it leaves his premises until clearance from customs at the point of entry into the country of installation.

2.13    Warranty

The contractor shall specify the period of the warranty together with a list of item covered under the warranty.

2.14    Service by Others

The following services after delivery of the equipment will be carried out by others:

1.    Clearance of the equipment through Customs;
2.    Transport to a location specified by the Purchaser, including insurance;
3.    Storage prior to erection if necessary;
4.    Construction of foundations;
5.    Erection snd setting to work of the equipment;
6.    Operation of the equipment; and
7.    Routine maintenance (as distinct from any repairs or maintenance required under the terms of the Warranty).


3.    QUESTIONNAIRE FOR TENDERERS

Tenderers are asked to supply the following information to demonstrate their ability to meet the requirements of the project. All information will remain confidential.


3.1    General Information

3.1.1  Name of Tenderer

       Individual Contact ...................................................................

3.1.2  Address ............................................................................
       ................................... Tel ................. Telex ................

3.1.3  Legal status (e.g. limited company) .................................................

3.1.4  Country in which registered ........................................................

3.1.5  Total number of employees ..........................................................


3.2    Associated or Subsidiary Companies

3.2.1  Briefly describe relationship with associated or subsidiary companies.

3.2.1  Overseas manufacturing/assembly subsidiaries (location and scope)

3.2.3  Licensees overseas for products (location, products


3.3    Experience of Tenderer

4.3.1  Number of years of experience with photovoltaic (PV) systems

4.3.2  Products developed

Experience of developing PV water pumping systems:

4.3.3  Number of years involved .................................................................

4.3.4  Number of systems installed worldwide .........................................................

4.3.5  Number of systems installed in developing countries ....................................

4.3.6  Number of systems now operational .........................................................

4.3.7  Special feature(s) of systems developed ......................................................

       ....................................................................................................
4.3.8  List of attached literature on systems currently available.

4.4    Source of Supply

4.4.1  Items manufactured by Contractor

4.4.2  Items bought in from Suppliers

3.5    Maintenance Requirements (detail and frequency)

       PV Array

       Controllers, batteries, switch gear, monitoring devices

       Motor

       Pump

       Pipework and Ancillaries


3.6    Spare Parts and Tools

       Tenderer to list all spare parts supplied as required to last for 10,000 hours of operation.

       Tenderer to  list all tools or equipment supplied for erecting, operating, maintaining and
       repairing the equipment.


3.7    After Sales Service

       Tenderers to list names, addresses, telex and telephone number of persons and organisations
       who may be contacted for advice during the period of installation and operation of the
       equipment:

4.    <u>PRICE AND DELIVERY</u>

Terms of Payment:

| Item | Description | Currency | Price |
|---|---|---|---|
| 1. | Design, manufacture and works testing of complete solar powered pumping system, including packing ready for despatch. | | |
| 2. | Transportation (from place of manufacture to point of entry) of complete pumping system, including insurance. | | |
| 3. | Other | | ---------- |
| | Total Contract Price | | ---------- |
| 4. | Spare parts | | |

    (a)  Photovoltaic Module
    (b)  Motor Unit
    (c)  Pump Unit
    (d)  Pump Impeller
    (e)  Motor Brushes
    (f)  Motor Bearings
    (g)  Pump Seals
    (h)  Controller etc.
    (i)  Others

Delivery of complete pumping system to be ........... weeks from receipt of order.

<u>APPENDIX 7.  GLOSSARY AND LIST OF SYMBOLS USED</u>

<u>Glossary</u>

**Annual equivalent life cycle cost (ALCC)**   The total life time costs of a pumping system expressed as a sum of annual payments.

**Clearness index**   The ratio of global solar irradiation to extraterrestrial solar irradiation.

**Design month**   For the purposes of sizing a solar, wind,  animal or hand pump it is convenient to choose  a  'worst  month' for which the pump must provide the water requirement.   This  month  is termed the design month.

**Diffuse radiation**   Solar radiation scattered by the atmosphere.

**Direct radiation**   Solar radiation transmitted directly through the atmosphere.

**Dynamic  head  or  Driving head**   The head loss in pipes caused by the flow of  water  through  the pipes.

**Effective rainfall**   The rainfall that makes a contribution to the crop water requirements.

**Efficiencies**

   The **daily energy efficiency**  (of the subsystem), is the ratio of hydraulic energy output over a day to electrical energy output of the PV array during the day.

   The **Power efficiency**  (of the subsystem),  is the ratio of hydraulic power output to electrical power output at any instant in time.

   The **volumetric efficiency**  of the storage and distribution system is the ratio of volume  of water delivered to the point of use to volume of water pumped.

   The **field application efficiency**  is the ratio of useful water taken up by the crop to  water delivered to the field.

   The **water conveyance efficiency**  is the ratio of water delivered to the  field  to  water provided at the outlet of the pump or storage tank/where installed.

**Energy equivalent**   The product of water requirements and total system head, expressed in $m^4$.

**Evapotranspiration**   Loss of water through the leaves of a crop which leads to the water requirements for the crop.

**Extra-terrestrial irradiation**   The solar energy received outside the earth's atmosphere.

**Field capacity**   The maximum amount of water held in the soil that is useful to the crop.

**Fill  factor**   The ratio of maximum power output of a PV cell under reference conditions  to  the product of open circuit voltage and short circuit current/under the same conditions.

**Global  irradiance**   The  sum of diffuse and direct solar irradiance incident  on  a  horizontal surface.

**Hydraulic energy**   The energy necessary to lift water.

**Impedance  matching**   The  process  of matching the output of one device to the input  of  another device such that there is a maximum transfer of power between the two.

**Inverter**   An  electronic  device for converting direct current  (d.c.)  to  alternating  current (a.c.).

**Life Cycle Costs (LCC)**   The lifetime costs associated with a pumping system expressed in terms of today's money.

**Maximum Power Point Tracker (MPPT)**   Impedance matching electronics used to hold the output of the PV array at its maximum value.

**Open circuit voltage** of a PV cell, module, or array is the voltage measured at the terminals when there is no electrical load.

**Packing factor** The ratio of cell area to gross area for a PV array.

**Peak demand factor (PDF)** The ratio of peak monthly water requirements in $m^3$ per day to average annual water requirements in $m^3$ per day.

**Permanent wilting point** The quality of water held in soil, below which the crop dies.

**Power conditioning equipment** Used to match the output of the PV array to the motor to obtain maximum power transfer.

**Present Worth** The value of a future cost or benefit expressed in present day money.

**Prime mover** The power source for a pumping system.

**PV or Solar Cell** A semi conductor device which can convert solar radiation directly into electricity.

**PV Module** The smallest complete environmentally protected assembly of interconnected solar cells.

**PV array** An assembly of PV modules together with the support structure.

**Peak watts (Wp)** The output of a PV module or array under reference conditions.

**Short circuit current** - of a PV cell, module or array is the current that flows when the output terminals of the device are joined together.

**Solar irradiance** The power received per unit area from the sun.

**Solar irradiation** (or insolation) The energy received per unit area from the sun in a specified time period. In this handbook, the time period is generally taken to be a day and the solar irradiation is expressed in MJ per $m^2$ per day.

**Static head** The vertical height through which a given quantity of water is lifted.

**Sub-system** The components of a solar pump that convert the electrical output of the PV array to useful hydraulic energy.

**Swept rotor area** The area swept by the blades of a windpump ($= \pi \times radius^2$).

**Threshold irradiation** The solar irradiation at which a solar pump commences or ceases to pump water.

**Tilt factor** The ratio of solar irradiation incident on a tilted PV array to global irradiation.

**Water conveyance network** - used to transport water from the pump (or storage tank) to the field.

<u>List</u> <u>of</u> <u>Symbols</u> <u>Used</u>

| | | |
|---|---|---|
| Ca | cost occuring each year | ($) |
| Cr | single cost occuring in the future | ($) |
| E | energy | (MJ) |
| g | gravitational acceleration | $(m/s^2)$ |
| h | head | (m) |
| H | solar irradiation | $(MJ/m^2)$ |
| H* | effective solar irradiation | $(MJ/m^2)$ |
| Pa, Pr | present worth factors (see Appendix 2) | |
| Q | water flow rate | (lit/sec) |
| U | wind speed | (m/s) |
| U* | effective wind speed | (m/s) |
| V | volume of water | $(m^3)$ |

www.ingramcontent.com/pod-product-compliance
Ingram Content Group UK Ltd.
Pitfield, Milton Keynes, MK11 3LW, UK
UKHW050009160726
7214IPUK00019B/349